GW00497774

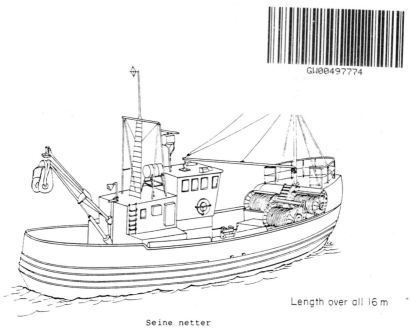

Length over all 16 m

Seine netter

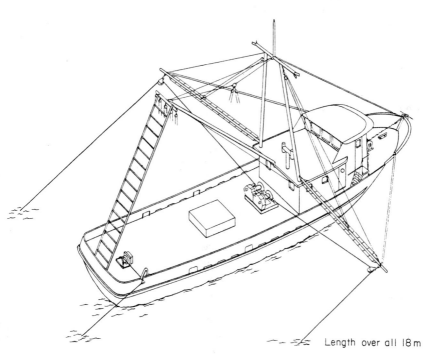

Length over all 18 m

Outrigger trawler

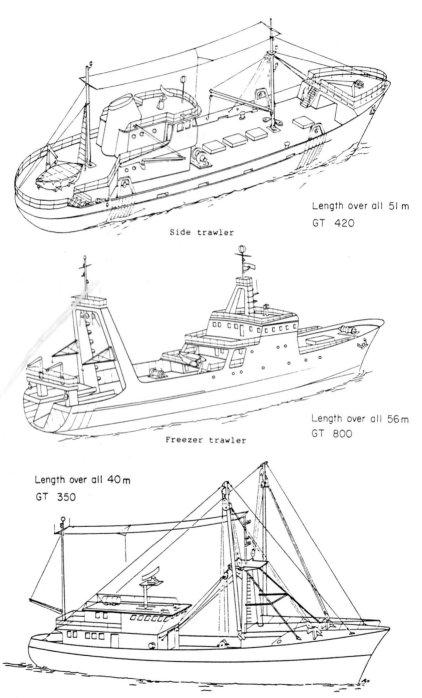

Length over all 51 m
GT 420

Side trawler

Length over all 56 m
GT 800

Freezer trawler

Length over all 40 m
GT 350

Beam trawler

Fishermen's Handbook

Edited by
Rear Admiral C R P C Branson CBE

for the
British Marine Mutual Insurance Association Limited

Published by
Fishing News Books Ltd
Farnham, Surrey, England

© British Marine Mutual Insurance Association Limited

First published 1980
Second edition 1987

British Library CIP Data

Fisherman's handbook.—2nd ed.
1. Seamanship 2. Fishing boats
I. Branson, C. R. P. C. II. British Marine
Mutual Insurance Association Limited
623.88′28   VK541

ISBN 0 85238 143 3 2nd edition
(ISBN 0 85238 106 9 1st edition)

Cover picture *Armana*
Courtesy of J. Marr & Son Ltd

Typeset by Mathematical Composition Setters Ltd., Ivy Street, Salisbury

Printed in Great Britain by Adlard & Son Ltd
Garden City Press, Dorking, Surrey

Contents

List of illustrations

8

Acknowledgments

I wish to express my gratitude for all the help I have received in preparing this revised edition; in particular to Charles Drever, MBE, a renowned skipper with wide experience of successful fishing for J Marr and Son, one of the leading trawler owners in Britain over the years. I have valued both his general advice and also his particular guidance on problems met on the fishing grounds.

Captain J Witty of the School of Fisheries Studies in the Humberside College of Higher Education has most willingly advised me over those aspects of this book which provide an introduction into a skipper's responsibilities; and has kindly assisted me in revising various texts.

Captain D J Bray (MNI) of the Department of Maritime Studies at the Lowestoft College of Further Education, has permitted me to draw on his clearly and simply expressed booklet 'Fishing Vessel Stability' which was originally prepared for a special course at that college following the foundering of a local fishing vessel from loss of stability.

The Department of Transport Marine Division has as always been helpful with information concerning safety and survival at sea.

The publishers of Reed's and Brown's Nautical Almanacs have allowed me to reproduce certain tables I believe to be of particular help to fishermen and which are included here for the first time.

Information on the international maritime buoyage system is reproduced by courtesy of the IALA

Finally I wish to record my thanks to the publishers who have throughout patiently assisted me in preparing this revised edition, not only with their advice and careful attention to detail but also by creating a specially pleasant and hospitable atmosphere when working in their home.

Rear Admiral C R P C Branson CBE

Preface to second edition

Since the last edition of the Fishermen's Handbook there have been substantial changes in the International Regulations for the Prevention of Collisions at Sea, 1972. These changes came into effect from 1st June, 1983, and are incorporated in this new edition. There has likewise been further progress in implementing the International Buoyage System; and this is also reflected.

More generally, fishermen should be aware of the 'IMO-Torremolinos International Convention for the Safety of Fishing Vessels, 1977'. Whilst the Convention applies specifically to new fishing vessels of 24 metres in length or over it affords useful guidance for owners of smaller fishing vessels. Indeed the Convention invites those Governments, which are signatories to it, to address themselves to smaller vessels as well as those over 24 metres in framing their national maritime regulations.

This edition also takes into account the fact that there already exist publications, such as Reed's Nautical Almanac, from which a skipper wishing to get a position line from the sextant observation of a heavenly body will find, in addition to the necessary almanac and tables, instruction in how to use them. The previous chapter on astronomical navigation has therefore been omitted. Likewise the chapter on Fishing Operations does not attempt to describe all the various methods and gear used in fishing on which very full literature already exists. Instead it deals with some of the hazards likely to be experienced in the more common modes, such as bottom trawling, purse seining, scallop dredging, gill netting and longlining.

It is also appreciated that handy references are available on national regulations and other local information needed by fishermen; this handbook therefore concentrates on the universal subjects of interest to them. A list of useful publications is included.

Successful commercial sea fishing is a demanding art and the sea itself an unforgiving taskmaster. Few men can have learned all the lessons it teaches at first hand so that even an experienced seaman should not be ashamed to seek advice. Certainly no fisherman can afford to neglect basic rules in meeting his responsibilities for the safety and efficiency of the vessel in which he sails. He is assisted by increasingly sophisticated technical aids: yet the traditional qualities of a sound knowledge of the sea and good judgment are still required as are a proper understanding of seamanship and navigation.

The aim of this book is to assist the aspiring fisherman in acquiring nautical skills ultimately required of a skipper and in so doing make a contribution to safety at sea in the exacting and changing world of deep-sea fishing.

Rear Admiral C R P C Branson CBE
British Marine Mutual Insurance Association Limited
Walsingham House
35 Seething Lane
London EC 3N 4DQ

Part I – Management

1 The fishing vessel

Commercial fishing vessels are subject to the rules and regulations of their country of registry which in turn take account of International Maritime Organisation (IMO) Regulations. A skipper should be aware of such regulations and the responsibilities thereby imposed on him. In the UK a fishing vessel certificate permitting it to operate is dependent on meeting various routine survey criteria concerning hull and machinery, stability, lifesaving and fire fighting equipment and radio.

The certificate of registry gives proof of ownership of the vessel, as well as details concerning its principal characteristics, overall length, register length, tonnage and date of build. It must at all times be carried on board. In the UK it remains valid for five years. Any alteration to the vessel affecting the measurements specified must be reported to the Registrar concerned, together with the certificate, for any necessary amendment: likewise any change of ownership. The vessel's registry is cancelled and the certificate given up when it is lost (including being deemed constructively lost by insurers), broken up or ceasing to be employed in commercial fishing.

Insurance

All prudent owners effect the most comprehensive insurance cover available to meet the vessel's operational requirements, including their liabilities to third parties and to crew. Broadly speaking these are covered by policies concerning hull and machinery, and protection and indemnity. Such policies require the vessel to be maintained and operated to specified standards and rules, and impose direct responsibilities on owners in ensuring compliance. A brief outline of the fishing vessel policy issued by the British Marine Mutual Insurance Association Limited is included at the back of this book. In addition to the requirements of national regulations concerning changes to the vessel and damage or other accidents experienced, either by the vessel or crew, it is a requirement of insurance policies that all such matters, including those caused to third parties, should immediately be reported to the vessel's insurers. The skipper, therefore, has the responsibility of noting all such incidents in the log and making such further detailed reports as will be needed for a proper assessment of the circumstances of the case and any consequential action needed. For example, following a grounding, insurers may require a vessel to be docked or slipped for a bottom survey and certificate of seaworthiness to be given.

Lloyds Register

Owners of the larger fishing vessels often elect to have them entered on the Lloyd's Register or with another classification society. This ensures that the construction and hull, machinery and equipment maintenance are to internationally recognised standards. This is also of particular assistance when repairs are effected following damage to hull or machinery.

Ship knowledge

The golden rule for all seamen is 'know your ship'. In particular it is too late to trace pumping and firemain systems once confronted with an emergency, which may occur in darkness. It is the responsibility of the skipper to ensure that every member of the crew should, on joining, familiarise himself with all the vessel's systems which may have to be operated in an emergency, as well as those used in steaming and in fishing operations. It should always be remembered that, unlike most ships, a fishing vessel's watertight integrity is constantly breached at sea by opening a hatch or hatches to stow the catch; this is often done in quite brisk weather and can lead to considerable amounts of sea water entering fish holds or gutting spaces; in the latter, running water is also used in cleaning fish. Fishing operations therefore entail much use of pumps and careful control of their valve systems.

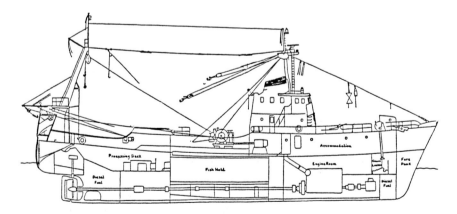

Fig 1.1 Profile of typical middle water trawler

General arrangement

The modern stern fishing middle water trawler is divided into various water-tight compartments each separated by a watertight bulkhead sometimes fitted with watertight access doors.

The forward end of the vessel incorporates a fore peak tank and forecastle arrangement with anchor cable locker.

The main compartments comprise the engine room and machinery space, propeller shaft tunnel, fish hold, fish processing space and after peak arrangement with ballast tanks and steering gear compartments. The layout of these compartments varies in the smaller vessels.

Machinery

The engine room/machinery compartment may be situated at the forward part of the vessel, with the main engine coupled to a reduction gearbox and with the propeller shaft extending via a tunnel compartment usually situated below the fish hold and processing deck compartments.

The main engine reduction gearbox may incorporate a reversing arrangement for fixed pitch propellers or may incorporate a control unit for variable pitch propellers.

The main engine is usually fitted with a forward end power take-off system incorporating drives for hydraulic machinery pumps and drives for electrical alternators.

An auxiliary engine is usually provided and coupled to an electrical alternator, and standby hydraulic system pumps.

Electrical power for the vessel is generated by either the main engine driven alternator or the auxiliary engine driven alternator and electrical power is used for driving auxiliary pumps such as bilge ballast, fuel transfer, engine standby pumps, lubricating oil pumps and cooling water pumps, and air compressors *etc*.

Pumping arrangements

Each watertight compartment within the vessel's construction is fitted with bilge pumping facilities usually arranged from the bilge main system.

The machinery space and propeller shaft tunnels can have a separate bilge direct suction system which allows the bilge pumps to take direct suction from these compartments without the necessity to connect the bilge pump to the bilge main.

Bilge pumping can be carried out by the use of two or sometimes three general service bilge and ballast pumps, at least one of which is independent of the main machinery. These pumps can be used variously for bilge pumping, ballast pumping and fire main functions by means of selection valves incorporated in a valve chest manifold.

These pumps may also be fitted with a sea suction connection valve which incorporates the use of non-return valves. However, it is emphasised that great care should be taken in the operation of sea suction valves and valve chest manifolds together with various compartment valves in order to avoid the possibility of flooding back through the system to the compartments via partially opened or defective non-return valves.

Bilge suction wells and bilge suction strainers should be kept clear of debris at all times in order to prevent any pumping problems. Particular attention should be given to areas where water and debris may accumulate, such as fishroom hold compartments and fish processing deck compartments.

The discharge of oily water residues from the bilges of machinery spaces is strictly controlled by the International regulations on Oil Pollution Prevention (MARPOL) and all pumping of dirty oil and oily water mixtures from machinery spaces should be monitored and a record of same be kept onboard the vessel.

Stability

Information regarding the vessel's stability is required by law to be kept onboard the vessel as follows:

1 Trim and stability booklet – This provides hydrostatic information which may be in the form of a graph or table, together with cross curves of stability and approximately six conditions of loading with regulations and minimum stability requirements. This information will also show maximum deck loadings for each hold and/or deck and tank capacities in terms of fuel, fresh water, and ballast water, including free surface moments.
2 Tank capacity plan; showing positions and quantities held by each tank.
3 General arrangement plan, showing an overall disposition of compartments, accommodation, and location of masts and fittings.
4 Docking plan; this shows the shape of the vessel's hull under the water line and includes any protrusions such as transducer fairings, twin propellers, which may have to be considered when dry docking.
5 Tank calibration tables, indicating tank capacities.

Ship construction and other definitions

Breasthook. A triangular plate bracket joining the port and starboard stringers to the stem.
Bracket. A plate used to connect rigidly two or more structural parts, *eg* deck beam to frame, bulkhead stiffener to frame.
Collision bulkhead or forepeak bulkhead. The foremost main transverse watertight bulkhead extending from the bottom of the hold up to the freeboard

deck and strong enough to withstand water pressure in the event of the fore-peak being open to the sea.

Deck stringer. The strake of plating on a deck immediately next to the hull.

Doublers. These are plates fitted inside or outside of and touching against another to give extra local strength or stiffness.

Flare. The spreading outwards of the hull form from the central vertical plane usually referred to as the forebody.

Forepeak. The watertight compartment at the extreme fore end (sometimes used for trimming purposes).

Frame. The transverse or longitudinal members of the hull structure stiffening the shell plating.

Freeing port. An opening in the lower part of the bulwark plating for draining off water, usually fitted with a hinged door or safety bars.

Intercostal. Made in separate parts, between frames, floors or beams.

Keel. The main ship's longitudinal member consisting of either a horizontal plate or a vertical bar fitted on the centreline of the bottom shell.

Longitudinals. Fore and aft sections or plates connected to the inner bottom, side shell or decks.

Margin plate. The outboard strake of the double bottom plating.

Pintles. The pins or bolts that connect the rudder to the gudgeons on the sternpost.

Scarph. A connection made between two pieces by tapering their ends to finish the joint in the same breadth and depth as the original pieces.

Sheerstrake. The strake of plating at the strength deck level.

Stiffener. A section, angle, tee, channel, bulb, trough or swedge on a panel of plating to impart stiffness.

Stringers. Fore and aft girders at the side shell; also the outboard strake of deck plating.

Transom. Square ended stern.

Tumblehome. The inward slope of a ships side generally above the waterline.

After perpendicular (AP) This line is perpendicular to the water line as defined, and positioned at the after side of the rudder post. If there is no rudder post the centre line of the rudder stock is taken to be the after perpendicular.

Forward perpendicular (FP) This is a vertical line passing through the intersection of the water line as defined, with the fore edge of the stem (*ie* stem line).

Length between perpendicular (LBP) This is the length as measured from the forward edge of the stem to the after edge of rudder post at the level of the waterline as defined.

Overall length. The greatest length of the vessel from the extreme forward end to the extreme after end.

Breadth moulded. The greatest breadth of the vessel to the outside of the frames.

Breadth extreme. The greatest breadth of the vessel to the outside of the shell.

Depth moulded. Measured from the top of the keel plate to the top of the uppermost continuous tier of beams at the ship side, this being measured at the middle length of the vessel.

Displacement. The mass of the volume of water dislodged by the vessel in any loaded condition.

Deadweight. The carrying capacity of the ship and includes fuel, feed water, fresh water, stores, crew, their effects and cargo.

Lightweight. The value:- loaded displacement – deadweight.

Amidships. The point mid way between AP and FP

Midship section. The transverse section of the ship amidships.

Trim. The difference between the draft forward and the draft aft. If the draft aft is greater than draft forward, then the vessel is said to have a trim by the stern; the reverse is called trim by the head.

Sheer. The curvature given to the decks in the longitudinal direction, and is measured at any point along the length by the difference in height, at the deck side, at that point and the height at the deck side amidships. The amount of sheer forward is usually twice the sheer aft, this being called 'Standard Sheer'.

Camber. The curvature given to the decks across the ship (transverse, athwartships) and is the difference in height measured at the centre line and that measured at the ship side.

2 Going to sea and return – maintenance

Before proceeding to sea a skipper must verify that the material state of the vessel is not impaired by any defects and that equipment meets statutory regulations. This entails a thorough check of all systems. Apart from the main engine and its emergency stop and emergency fuel shut off, where so fitted, auxiliary engines and associated pumping systems will also have been tested, likewise all forms of steering control (a mandatory requirement) and the various alarm systems fitted. All bilges will have been pumped out and strainers cleaned, particular attention being paid to 'slush wells' in the fish hold. The gyro will have been run up several hours before sailing and its accuracy checked; likewise radar which, together with the echo sounder, must be available on sailing. All main and auxiliary navigation lights should have been checked and cleaned where necessary and the siren and auxiliary sound equipment tested.

The skipper will have ensured that the correct quantity of fuel is embarked, taking into account any restrictions imposed to meet stability criteria for various kinds of fishing, eg bulk-fishing of pelagic species. He will have ensured that required stores, provisions and fresh water are embarked. In particular he will verify that his chart outfit and publications adequately cover the intended area of the fishing trip and are corrected up to date by checking that he has the latest number of Admiralty Notices to Mariners. His passage plan to the fishing grounds, however often employed, must have taken adverse weather forecasts into account, particularly ice conditions where these apply.

The crew engaged, which must include men holding qualifications required by national regulations, will have been entered in the Official Log in accordance with such regulations.

Immediately prior to sailing a skipper will clear his departure with customs or dock authorities as required locally, in particular checking tidal conditions affecting passage out of basin lock gates and shallow estuaries.

Either before sailing or shortly afterwards the skipper will brief and exercise the crew as to their stations and responsibilities in various emergencies, including the use of appropriate fire fighting equipment.

Return to harbour

In normal circumstances a skipper will radio his intended time of arrival at a port to dock and customs authorities, having previously checked that tidal conditions are suitable to reach intended berth. He will also request medical clearance if the vessel has been to a foreign port during the last trip. Before entering the harbour approaches he may consider it prudent to test astern power if this has not been used for some time.

Maintenance

Ultimate checks on the overall standards of a fishing vessel's hull, machinery and equipment are afforded by the official surveys required by national administrations. In the case of the larger fishing vessels entered in Lloyd's Register or other classification society these are supplemented by the survey rules of the Register. Marine insurance companies will also require a vessel to be surveyed before accepting it 'on risk' and following incidents of suspected or actual damage.

But proper maintenance of a vessel cannot be left at that: it is a continuous process determined on the one hand by the makers' recommendations concerning machinery and other installed equipment, as for example the engine running hours between top overhauls; and, on the other, by common sense. For instance, following the presence of plastic bags, synthetic twines or ropes in the water around his vessel, a prudent skipper will ensure none has been drawn into a sea suction. He will likewise have a diver examine the propeller if he suspects it has been fouled.

It is a false economy to forego routine overhauls when things seem to be working satisfactorily. Should important machinery subsequently break down and the cause be attributed to 'fair wear and tear' the expenses of defect rectification may fall to the owner and not be a claim on the vessels' insurance policy. Breakdowns at sea entail loss of fishing with a full crew onboard whereas routine harbour maintenance can usually be co-ordinated and planned to suit the skipper's general operating arrangements. Moreover relying on 'breakdown maintenance' can often lead to an incorrect assessment of basic cause and so to further expenses when initial attempts to rectify a defect prove unsuccessful.

3 Fishing vessel stability

The word 'stability' means the ability of a vessel to return upright after being forcibly heeled by an outside force. There are many factors affecting the stability of a vessel, and a knowledge of these factors and their effects is essential for the skipper of a fishing vessel.

Transverse stability

Dynamic stability is the energy which will return a ship to its normal equilibrium when it is moved by an outside force such as wind or wave motion. When a vessel is placed in water it will displace a volume of water equal to that of its own weight. If the weight of the vessel and its contents exceeds the weight of the volume of water displaced, it will sink.

The buoyant volume of a ship is that part of the vessel which is enclosed and is watertight. This is the total buoyancy force, and that part which is normally above the water plane is known as the reserve buoyancy.

To be able to understand stability, the fisherman must understand the following terms: the centre of gravity, the centre of buoyancy and the metacentric height.

Basic principles

The stability of a vessel depends upon the positions of its centre of buoyancy and centre of gravity.

The centre of gravity (G) is the position in the ship (by design on the centreline) where all the weight in the vessel can be said to act. This position does not normally move. It will only move if weight is added to or removed from the vessel, or if weight onboard is moved.

The centre of buoyancy (B) is best described as the central point of the underwater volume of the vessel. It is the point through which the forces of buoyancy act vertically upwards. The position of the centre of buoyancy will change continually as the vessel moves in a seaway, or as her draught, heel or trim change. The force of buoyancy is equal to the amount of water displaced. For a vessel to float upright in still water she must displace her own weight of water, and her centres of gravity and buoyancy must lie in the same vertical line as shown in *Fig 3.1*.

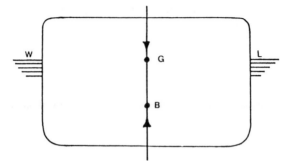

Fig 3.1 Centre of gravity and buoyancy

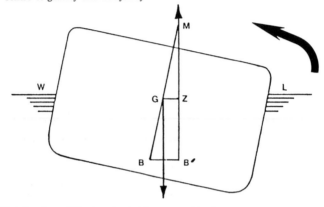

Fig 3.2 Righting lever GZ when heeled by an outside force

As a vessel starts to heel from the influence of an outside force the underwater profile of the hull changes and the centre of buoyancy B moves out to the low side to B' forming a 'Righting Lever' (GZ) as shown in *Fig 3.2*.

The metacentre is the point of intersection of verticals drawn through the position of B when the vessel is upright and B' when heeled to relatively small angles. The distance GM is known as the metacentric height.

The metacentre M may be regarded as a fixed point but only at small angles of heel, and for a particular draught. At large angles of heel the draught and underwater shape change quickly and other more complicated calculations have to be made. If the designer arranges the hull dimensions so that at small angles of heel there is a good safe margin of stability and there is a reasonable stability factor for different but normal loading conditions, then it may be assumed that the conventional form of vessel will have sufficient transverse stability to correct heel when rolling in heavy weather.

The distance of G below M is thus a measure of the initial stability of the vessel at small angles of heel; and for a vessel to be stable G must be below M. If G moves above M the GM becomes negative and the vessel unstable.

Knowing the length of GM and an angle of heel it is possible to calculate GZ, the 'righting lever' at such an angle. Multiplying GZ by the displacement gives the moment of force which will bring the ship back to the upright when she has been heeled by an outside force.

The righting lever

Although GM is a good guide to initial stability one must also consider the amount of righting lever the vessel has at larger angles of heel. Looking at the inclined vessel in *Fig 3.2* the line GZ (the horizontal separation between B and G) is the length of the righting lever, and it will vary with the angle of heel. If a vessel is heeled to increasingly greater angles, we see in *Fig 3.3* that the righting lever increases up to a certain angle, and then decreases beyond it.

At a certain angle of heel the righting lever diminishes to nothing, and this is known as the angle of vanishing stability. If the vessel is heeled further than this, righting levers become negative, *ie* they become heeling levers, and the vessel will not recover, but capsize.

It is important to note that the amount of stability a vessel has at any given angle of heel is proportional to the length of its righting lever at that angle. A graph may be drawn showing length of righting lever against angles of heel. This curve is known as the GZ curve, or the statical stability curve. *Fig 3.4* shows how the GZ curve is determined. Such a curve shows us the angle at which occurs the maximum righting lever, the angle of vanishing stability and the range of stability.

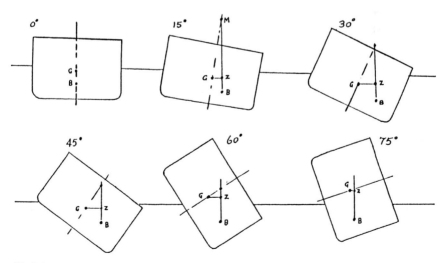

Fig 3.3 Righting lever at angles of heel

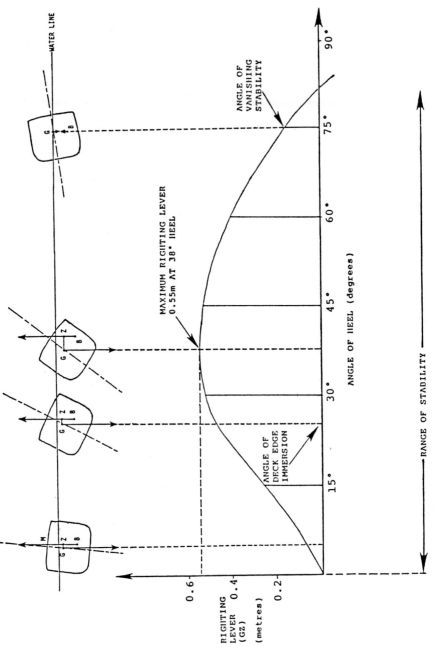

Fig 3.4 Typical statical stability curve with angles of heel above

On the curve shown in *Fig 3.4*, the maximum righting lever occurs at approximately 38° angle of heel. It must be realised that although the vessel has the ability to right itself beyond this angle, its power of recovery progressively diminishes beyond this angle.

The *range of stability* is the range of angles at which the vessel has a positive righting lever. The vessel will recover if inclined to any angle within the range of stability, but will capsize if heeled beyond her range of stability.

The overall stability of a ship is proportional to the total area under the curve. The greater this area then the greater the vessel's ability to recover from heeling.

Stiff and tender vessels

If a vessel has a large transverse metacentric height (GM), she will develop a short but rapid roll and in very bad weather it may be described as violent. Such a ship is said to be stiff and because of the rapid roll may damage herself by shifting cargo, machinery, *etc*, or crew may suffer injury.

A vessel with a small GM will develop a long slow transverse motion and is said to be tender. The ideal condition is somewhere in between with a reasonable stability factor that would make the ship neither stiff nor tender.

Factors affecting stability

So much for the basic concepts. Let us now see how the stability can be affected adversely. When a vessel sails and she is fully bunkered and stored, the position of M the metacentric, which has been determined by the designer for that draught, is fixed. It is fixed for various draughts and M will be tabulated and often shown as a diagram which will show BM, *ie* the distance of the transverse metacentre M from B the centre of buoyancy. Its height will also be shown as a distance from the keel.

G, the centre of gravity, can, however, be altered by moving stores, fuel or ballast. Traditionally, trawlermen would sail from their home port bound for distant waters and they would be at sea for some 20 days. A large sized distant water vessel would carry some 180 tons of fuel, several tons of fresh water and stores. During the three week period fuel, water and stores would be consumed and the vessel would return with about 50 tons of fuel on board.

But during this period the trawler would be progressively catching fish, so balancing out the consumption of fuel, *etc*. A catch of 2,000–2,500 kits of fish would equal 125–155 tons in weight, which would approximate with the stores and fuel used.

Traditionally, stability has not posed many problems in trawler operations other than in such exceptional circumstances, as:

— Heavy icing on deck and superstructure, when G would be raised until

there was little or no righting moment and positive stability. Draught would increase and freeboard reduce.
— The shipping of heavy and successive seas in abnormally bad weather. If the heavy volume of water on deck is not free to spill quickly from a well deck, it will have the effect of raising the vessel's G, increase the draught, and reduce the freeboard; the centre of gravity of the sea water will move from side to side if the vessel is rolling.

With the advent of the pelagic mode of fishing, circumstances affecting stability have been changed considerably and must be carefully considered.

The basic reasons for these changes are that pelagic fishing allows for a large weight of catch to be brought on board in a very short time. In the traditional demersal fishing it might take eleven days of continuous fishing to produce 130 tons of fish. This weight of fish may now be caught pelagically in two or three hauls and in a matter of hours. The factors adversely affecting stability are as follows:

— The extra weight of equipment, net drum and gear may add 10–20 tons of top weight to a vessel, depending on size of ship. This will raise G.
— As much as 30–40 tons of fish may be brought on deck in one haul in a stern trawler, depending on the size of ship. This will raise G.
— If the fish is bulk stowed and not divided properly it will move if the ship heels or rolls so that G will move proportionally.
— By reason of only having to be at sea for a very short time the vessel may only be lightly stored and bunkered. The double bottom tanks may be slack because of the minimum fuel on board. Some tanks may be empty. This will raise G.
— The cubic capacity required to stow one ton of demersal fish such as cod and haddock, iced and shelved is high. The cubic capacity to stow one ton of pelagic fish in bulk with the water element is low. By filling the fishroom and taking fish on deck, the vessel could be overloaded, with a consequent increase in draught and decrease of stability and freeboard.

If a side trawler lifts large weights of fish by derrick or gilson wire, then she will reduce her stability considerably, and will list heavily when the weight comes on the derrick head.

Providing that the skipper has an understanding of stability and the danger factors there is no reason why trawlers should not be able to catch and stow relatively large weights of fish. It is only prudent, however, to have new stability data drawn up by the shipbuilder or other authority showing GM for the new conditions. Various recommendations may be made for different ships, ie a ship may have to have permanent ballast in order to increase her GM when in light condition in order to offset a large weight of fish on deck. The prudent skipper will study the stability plans, curves and information, and will not overload his vessel.

Stability information, which includes typical conditions of loading, is provided for use by the skipper on board. This will allow him to acquire a general understanding as to how the stability of his vessel will fluctuate during a typical voyage and will also provide the basis for any adjustment he may wish to make for a special or particular condition of loading. An 'icing-up' weight will usually be given for trawlers likely to experience such conditions.

Before looking at plans or conditions of loading, the fisherman should understand that at this stage we have been looking at the transverse stability of a ship. Transverse stability is related directly to the vertical height of the centre of gravity measured from the keel. In ships' plans or tables it will be shown as VCG. Its height influences the speed and degree to which a ship will roll or heel. VCG will move up or down towards a weight taken on board, depending where the weight is placed. It will move when a weight is taken out of a ship, away from the position of the weight. It will move sideways towards a weight if the weight is loaded off centre. *Fig 3.5* shows the effect of picking up a bag of fish from over the side by derrick or gilson. It demonstrates that the centre of gravity moves not only vertically but horizontally towards the derrick head. If we suppose M to be fixed it will be seen that the value of GM and the righting lever has been considerably reduced. If the above effect is combined with an increased movement of G when the ship is listed owing to a slack tank as shown in *Fig 3.9* an aggravated situation will arise.

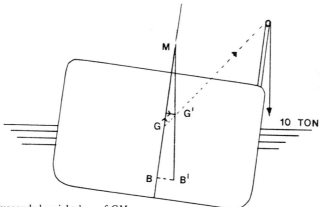

Fig 3.5 Suspended weight loss of GM

The centre of gravity is also measured longitudinally from the centre of flotation (CF). CF or tipping centre is the geometrical centre of the water plane about which the vessel will trim. It is important to the fisherman to know that the longitudinal centre of gravity will affect trim. When a heavy bag of fish is hauled on to the ramp of a stern trawler, the vessel will increase the mean draught because of the extra weight being taken on board. But because the weight is taken on to the ramp at the extreme distance from the

CF then a considerable change in trim will take place. The ship's draught will increase aft and decrease forward. In the ship's plans the longitudinal centre of gravity will be seen as LCG.

From all we have discussed it will be seen that the dangers facing the fisherman may be a complete loss of GM and/or GZ, the righting lever. The causes may now be separated into different categories for description, action and correction, *ie* list, loll, free surface, suspended weights.

List

List is caused by off centre loading. The ship's centre of gravity will move towards the weight in a ratio dependent upon the weight loaded or moved, its distance from the centre line, and the vessel's displacement. The list may be corrected by adjustment and movement of the weight or other weights about the centre line so that the vessel will return to an upright position. (See *Fig 3.6*).

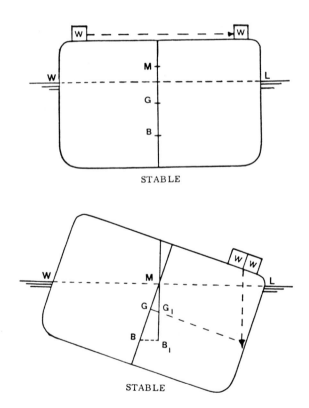

Fig 3.6 Effect of a moved solid weight creating a list.

If a list is not corrected, however, the listed position becomes the initial condition of stable equilibrium and the vessel will roll about this position. If we suppose that a trawler has an uncorrected list of $10°$ to starboard, then in the event of rolling due to bad weather, the vessel would roll symmetrically about the inclination of $10°$ to starboard: when rolling to port she would reach the vertical and when rolling to starboard she would be heeled $20°$ from the vertical. This would not be desirable and the list should be corrected by adjusting solid weights around the centre line of the ship.

Loll – the unstable condition

We have already noted the need for a vessel to be so designed as to have a good margin of stability for various but normal loading conditions to correct heel when rolling in heavy weather and ensure it does not roll to the angle of vanishing stability.

But if the centre of gravity of the vessel is allowed to rise (from excessive top weight, possibly in combination with too little bottom weight or the effect of free surface liquids in slack tanks or on deck) then the lengths of the righting levers are reduced at all angles of heel and the angle of vanishing stability is also reduced. While this may only be an awkward situation in harbour it is unacceptable at sea.

If G rises to the point where GM becomes zero, a situation of neutral equilibrium arises where the vessel will, in calm water, have no tendency to return upright from any heel angle. She will not capsize of her own accord but she will not resist heeling forces as there is no righting lever.

A further rise in G will make the vessel unstable. The GM is now negative, also the righting lever, causing the vessel to heel. She will not stay upright and is said to be in a state of loll.

The first indication of such a condition being reached is a tendency for the vessel to flop from side to side under any heeling forces, coming to rest, in calm water, at what is termed the angle of loll, where G and M are coincident. See *Fig 3.7*.

At any further angle of heel caused by an external force, the increased waterplane area, from increased breadth while mean draught remains constant, will lead to the position of the metacentre changing and M moving upwards, causing a positive righting lever to be formed, albeit a small one, returning the vessel to the angle of loll, about which the vessel therefore moves.

It will, however, be readily apparent that with the vessel already heeled at its angle of loll and having a much reduced righting lever, the angle of vanishing stability is much less and if occurring at sea the ship's condition is highly dangerous, easily leading to capsize.

By reducing the top weight and reducing free surface, G may be lowered. It is important that counter weights which would correct a list should not be

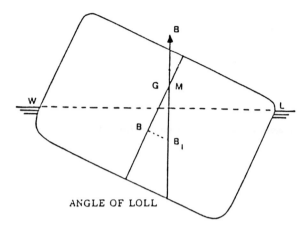

ANGLE OF LOLL

Fig 3.7 Angle of loll

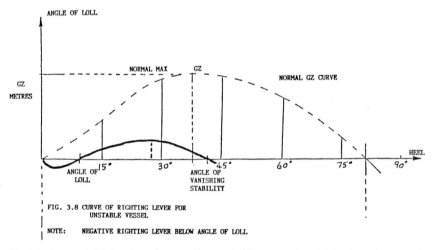

Fig 3.8 Curve of righting lever for unstable vessel. Note: negative righting lever below angle of loll.

used. In using counter measures such as filling a side tank on the high side or removing weight from the low side, it must be remembered that because of the low stability factor or wind/wave action, the vessel could easily roll to the other side and the momentum gained in the roll may take her past the angle of loll and she may not recover. G can be lowered by:

— Filling a small divided double bottom tank on the *low* side. If it is already slack, well and good. If empty, G will rise at first but as the tank becomes full G will fall. The angle of heel may now be a list which can be eliminated by filling the opposite tank on the high side.

— Jettison weights from the *high* side. if the loll has been caused by icing up on the trawler's masts, stays, *etc*, remove the ice beginning on the centre line, masts, *etc*, then on the high side, before starting on the low side.

In both cases as set out above, G is lowered and the angle of heel may at first increase. Remember that the waterplane area initially increases from the ship's vertical position as she heels.

At the beginning of this chapter, certain factors were set out, concerning the pelagic mode of fishing. These factors included heavy weights taken on deck with empty and/or slack bottom tanks. Any combination of these factors might lead to a loll condition. Beam seas would further aggravate the situation.

Free surface effect

Provided that a tank is at least 98% full of liquid, no movement of the liquid is possible, and the tank may be regarded as being filled with a solid material. When a quantity of liquid is drawn off, however, the stability of the ship is adversely affected by what is known as 'free surface effects'. Regardless of where the tank is situated within the ship, either high or low, there is a rise of G and a loss of GM depending on the length and breadth of the tank and the relative density of the liquid.

The liquid in an undivided tank running across the breadth of the ship and partially filled must be regarded as a weight which is free to move from side to side as the ship heels when steaming at sea. The G of the liquid moves from side to side and the G of the vessel consequently moves proportionally from side to side of the centre line. If we examine *Fig 3.9* it will be seen how a slack tank reduces GM.

If, however, the tank is divided at the centre line and we have two separate tanks, the loss of GM is reduced to 1/4 of that of the undivided tank, *ie* the square of the number of tanks. If the tank was divided longitudinally into four tanks the loss of GM would be reduced to 1/16th of the undivided tank. (See *Fig 3.9a, b*).

Figures regarding free surface effect for each tank are given in the ship's stability booklet. They are generally given in the form of 'free surface moments' which must be divided by the displacement of the ship to give the reduction of the metacentric height in metres. If more than one tank is slack the free surface moments for each slack tank must be added, then the total may be divided by the displacement to give the total reduction in GM due to free surface effects on board.

The GM of a vessel, not making any allowance for any free surface effect is known as 'solid GM'.

The GM of a vessel after allowing for all free surface effects on board is

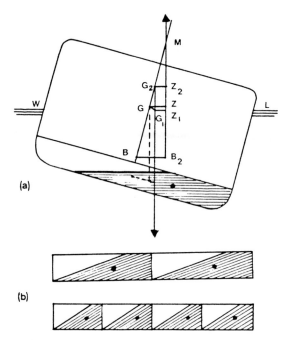

Fig 3.9(a) Loss of stability due to slack tanks
   (b) Reduction in stability loss with the greater number of tanks

known as the 'fluid GM' and this figure must be used as the effective GM.

Free surface effect may occur anywhere where there is a fluid mass. If a vessel takes a heavy sea on deck, then, in addition to the reduction in stability from all this added top weight, she has also gained an enormous free surface effect from the water on deck. If this cannot clear quickly the vessel may be overwhelmed by heavy seas. The skipper must ensure that his freeing ports are kept clear and in working order, such that deck water may readily escape.

Free surface may also occur in the fishroom, either from loose water, or fish. Loose water in the fishroom must be kept to a minimum, and the slush well kept clear.

Suspended weights

The principle of suspended weights is that when a weight is picked up by a derrick, the centre of gravity G of the weight moves immediately to the derrick head and this happens regardless of the position of the weight, *ie* whether it is just clear of the water or right up at the derrick head. Therefore, we have two stability factors to consider. (See *Fig 3.5*).

In a side trawler the weight of a bag of fish taken from the water will be imposed at the derrick head which is a considerable distance vertically from

G the ship's centre of gravity. The ship's G will be raised by the weight of the bag in a vertical direction.

The second stability factor will be that the ship's G will move sideways towards the weight and because the derrick head is outboard, the distance from the centre line will be considerable. Depending on the weight of the fish in the cod end a list will develop. Once again we must compare traditional fishing methods with the pelagic mode at present being used. If a side trawler fishing demersally picked up a cod end containing 100 baskets of fish, about 1¼ tons in weight, then picking this up from over the side by gilson or derrick would have little effect on stability.

But a large pelagic net with 15–20 tons of fish would very much affect the stability of a trawler in the 500 ton displacement range. Far better to pump the fish on board hydraulically from the net from its buoyant position alongside. It may be brailed on board as in seine net fishing but this would be a long and tedious operation.

(*Note*: A similar hazard is experienced if, for example, a scallop dredger fouls and hoists a length of old, cut telegraph cable)

Hints on practical stability for fishermen

A fairly safe estimate for GM which might be found useful is that the GM of any fishing vessel might be considered to be reasonable if it lies somewhere about 4% of the ship's breadth. If a table of a small, medium and large stern trawler is made up it will be seen that 3%–4% gives a reasonable margin of stability in a properly loaded vessel. It is a rule of thumb method only.

Trawler	Beam	%	GM
Small	6.01m or 20ft	4%	.24m or 0.8ft
Medium	8.5m or 27.89ft	4%	.34m or 1.12ft
Large	12.2m or 40ft	4%	.49m or 1.6ft

Freeboard and stability

The International Convention on the Safety of Fishing Vessels makes recommendations concerning stability criteria, safe working on deck and watertight integrity. In the UK these criteria include rules governing minimum freeboard heights, which takes the place of load line regulations.

It is most important to appreciate the effect of a reduction in freeboard upon the stability of a fishing vessel.

Fishermen must therefore always bear in mind that reduced freeboard means reduced stability.

36

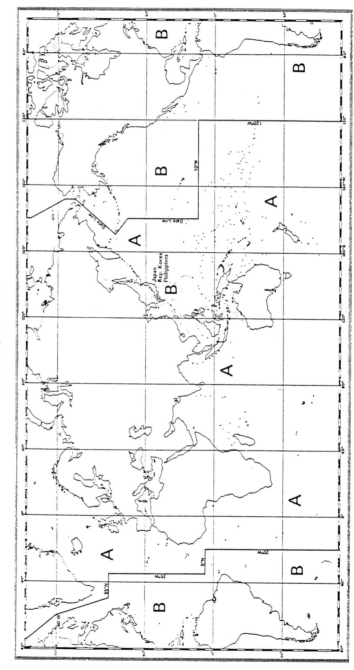

IALA MARITIME BUOYAGE SYSTEM
Buoyage Regions A and B

Fig 4.1 IALA Maritime Buoyage System – Regions A and B

Part II – Navigation

4 Maritime buoyage systems - International Association of Lighthouse Authorities (IALA)

At a conference convened by IALA in November 1980, with the assistance of IMCO (now IMO) and the International Hydrographic Organisation (IHO), the Lighthouse Authorities from 50 countries and the representatives of 9 International Organisations concerned with aids to navigation met, and agreed to adopt the rules of the new combined Cardinal and Lateral System. The boundaries of the buoyage regions were also decided and illustrated on the map annexed to the rules. *Fig 4.1*.

From this it will be seen that Region A includes Europe, Australia, New Zealand, Africa, the Gulf and some Asian countries.

Region B includes North, Central and South America, Japan, Korea and the Philippines.

The new system is already in use in all European waters except for certain areas of the Eastern Mediterranean. It is hoped the system will be virtually complete world-wide by 1990.

Meanwhile fishermen intending to navigate where introduction of the system is still in progress should consult Notices to Mariners and the latest charts for information.

The combined Cardinal and Lateral System (see *pages 313* to *316*)

The system applies to all fixed and floating marks, other than lighthouses, sector lights, leading lights and marks, light-vessels and lanbys. It serves to indicate:

The sides and centrelines of navigable channels; natural dangers and other obstructions such as wrecks (which are described as 'New Dangers' when newly discovered); areas in which navigation may be subject to regulation; or other features of importance to the mariner.

There are five types of marks in the system which may be used in any combination, each of the five types of marks have significant characteristics which depend on the following features:

By *night*: light, colour and rhythm

By *day*: colour, shape and topmark

Lateral marks differ between Buoyage Regions A and B as described later, whereas the other 4 types of mark are common to both regions.

Lateral marks

Lateral marks are generally used for well-defined channels; they indicate the port and starboard hand sides of the route to be followed, and are used in conjunction with a conventional direction of buoyage.

Following the sense of a 'conventional direction of buoyage', Lateral marks in Region A utilize red and green colours by day and night to denote the port and starboard sides of channels respectively. However, in Region B these colours are reversed with red to starboard and green to port.

A modified Lateral mark may be used at the point where a channel divides to distinguish the preferred channel, that is to say the primary route or channel which is so designated by an Authority.

Cardinal marks

Cardinal marks indicate that the deepest water in the area lies to the named side of the mark. This convention is necessary even though for example, a North mark may have navigable water not only to the North but also East and West of it. The mariner will know he is safe to the North, but must consult his chart for further guidance.

Cardinal marks do not have a distinctive shape but are normally pillar or spar. They are always painted in yellow and black horizontal bands and their distinctive double cone top-marks are always black.

An aide-memoire to their colouring is provided by regarding the topmarks as pointers to the positions of the black band(s):

Topmarks pointing upward: black band above yellow band

Topmarks pointing downward: black band below yellow band

Topmarks pointing away from each other: black bands above and below a yellow band

Topmarks pointing towards each other: black band with yellow bands above and below

Cardinal marks also have a special system of flashing white lights. The rhythms are basically all 'very quick' (VQ) or 'quick' (Q) flashing but broken into varying lengths of the flashing phase. 'Very quick flashing' is defined as a light flashing at a rate of either 120 or 100 flashes per minute, 'quick flashing' is a light flashing at either 60 or 50 flashes per minute.

The characters used for Cardinal marks will be seen to be as follows:

North: Continuous very quick flashing or quick flashing

East: Three 'very quick' or 'quick' flashes followed by darkness

South: Six 'very quick' or 'quick' flashes followed immediately by a long flash, then darkness

West: Nine 'very quick' or 'quick' flashes followed by darkness.

The concept of three, six, nine is easily remembered when one associates it with a clock face. The long flash, defined as a light appearance of not less than 2 seconds, is merely a device to ensure that three or nine 'very quick' or 'quick' flashes cannot be mistaken for six.

It will be observed that two other marks use white lights. Each has a distinctive light rhythm which cannot be confused with the 'very quick' or 'quick' flashing light of the Cardinal marks.

Fig 4.2 Conventional buoyage direction around the British Isles

Isolated danger mark

The isolated danger mark is placed on a danger of small area which has navigable water all around it. Distinctive double black spherical topmarks and group flashing (2) white lights, serve to associate isolated danger marks with Cardinal marks.

Safe water marks

The safe water mark has navigable water all around it but does not mark a danger. Safe water marks can be used, for example, as mid-channel or land-fall marks.

Safe water marks have an appearance quite different from danger marking buoys. They are spherical, or alternatively pillar or spar with a single red spherical topmark. They are the only type of mark to have vertical stripes (red and white). Their lights, if any, are white using isophase, occulting, one long flash or morse 'A' rhythms.

Special marks

Special marks are not primarily intended to assist navigation but are used to indicate a special area or feature whose nature may be apparent from reference to a chart or other nautical document.

Special marks are yellow. They may carry a yellow 'X' topmark, and any light used is also yellow. To avoid the possibility of confusion between yellow and white in poor visibility, the yellow lights of Special marks do not have any of the rhythms used for white lights.

Their shape will not conflict with that of navigational marks; this means, for example, that a special buoy located on the port hand side of a channel may be cylindrical, but will not be conical. Special marks may also be lettered or numbered to indicate their purpose.

New dangers

It should be specially noted that a 'new danger' which is one not yet shown in nautical documents, may be indicated by exactly duplicating the normal mark until the information is sufficiently promulgated. A 'new danger' mark may carry a Racon coded Morse 'D'.

Lateral marks

Definition of 'conventional direction of buoyage'

The 'conventional direction of buoyage', which must be indicated in appropriate nautical documents, may be either:

The general direction taken by the mariner when approaching a harbour, river, estuary or other waterway from seaward, or
The direction determined by the proper authority in consultation, where appropriate, with neighbouring countries. In principle it should follow a clockwise direction around landmasses.

Description of lateral marks used in region A

Port hand marks		Starboard hand marks	
Colour:	Red	Colour:	Green
Shape (buoys):	Cylindrical (can), pillar or spar	Shape (buoys):	Conical, pillar or spar
Topmark (if any):	Single red cylinder (can)	Topmark (if any):	Single green cone point upward
Light (when fitted):		Light (when fitted):	
Colour:	Red	Colour:	Green
Rhythm:	Any, other than that described in preferred channel below	Rhythm:	Any, other than that described in preferred channel below

At the point where a channel divides, when proceeding in the 'conventional direction of buoyage', a preferred channel may be indicated by a modified port or starboard lateral mark as follows:

Preferred channel to starboard:		Preferred channel to port:	
Colour:	Red with one broad green horizontal band	Colour:	Green with one broad red horizontal band
Shape (buoys):	Cylindrical (can), pillar or spar	Shape (buoys):	Conical, pillar or spar
Topmark (if any):	Single red cylinder (can)	Topmark (if any):	Single green cone, point upward
Light (when fitted):		Light (when fitted):	
Colour:	Red	Colour:	Green
Rhythm:	Composite group flashing (2 + 1)	Rhythm:	Composite group flashing (2 + 1)

Description of lateral marks used in region B

Port hand marks		Starboard hand marks	
Colour:	Green	Colour:	Red
Shape (buoys):	Cylindrical (can), pillar or spar	Shape (buoys):	Conical, pillar or spar
Topmark (if any):	Single green cylinder (can)	Topmark (if any):	Single red cone, point upward
Light (when fitted):		Light (when fitted):	
Colour:	Green	Colour:	Red
Rhythm:	Any, other than that described in preferred channnel below	Rhythm:	Any, other than that described in preferred channel below

At the point where a channel divides, when proceeding in the 'conventional direction of buoyage', a preferred channel may be indicated by a modified port or starboard lateral mark as follows:

Preferred channel to starboard:		*Preferred channel to port:*	
Colour:	Green with one broad red horizontal band	Colour:	Red with one broad green horizontal band
Shape (buoys):	Cylindrical (can), pillar or spar	Shape (buoys):	Conical, pillar or spar
Topmark (if any):	Single green cylinder (can)	Topmark (if any):	Single red cone, point upward
Light (when fitted):		Light (when fitted):	
Colour:	Green	Colour:	Red
Rhythm:	Composite group flashing (2 + 1)	Rhythm:	Composite group flashing (2 + 1)

General Rules for lateral marks

Shapes

Where lateral marks do not rely upon cylindrical (can) or conical buoy shapes for identification they should, where practicable, carry the appropriate topmark.

Numbering or lettering

If marks at the sides of a channel are numbered or lettered, the numbering or lettering shall follow the 'conventional direction of buoyage'.

Cardinal marks

Definition of cardinal quadrants and marks

The four quadrants (north, east, south and west) are bounded by the true bearings NW-NE, NE-SE, SE-SW, SW-NW, taken from the point of interest.

A cardinal mark is named after the quadrant in which it is placed.

The name of a cardinal mark indicates that it should be passed to the named side of the mark, *eg* pass North of a North mark *etc.*

Use of cardinal marks

A cardinal mark may be used, for example:

to indicate that the deepest water in that area is on the named side of the mark

to indicate the safe side on which to pass a danger

to draw attention to a feature in a channel such as a bend, a junction, a bifurcation or the end of a shoal.

Description of cardinal marks

North cardinal mark

Topmark:	2 black cones, one above the other, points upward
Colour:	Black above yellow
Shape:	Pillar or spar
Light (when fitted):	
Colour:	White
Rhythm:	VQ or Q

East cardinal mark

Topmark:	2 black cones, one above the other, base to base
Colour:	Black with a single broad horizontal yellow band
Shape:	Pillar or spar
Light (when fitted):	
Colour:	White
Rhythm:	VQ(3) every 5s or Q(3) every 10s

South cardinal mark

Topmark:	2 black cones, one above the other, points downward
Colour:	Yellow above black
Shape:	Pillar or spar
Light (when fitted):	
Colour:	White
Rhythm:	VQ(6) + Long flash every 10s or Q(6) + Long flash every 15s

West cardinal mark

Topmark:	2 black cones, one above the other, point to point
Colour:	Yellow with a single broad horizontal black band
Shape:	Pillar or spar
Light (when fitted):	
Colour:	White
Rhythm:	VQ(9) every 10s or Q(9) every 15s

Isolated danger marks

Definition of isolated danger marks

An isolated danger mark is a mark erected on, or moored on or above, an isolated danger which has navigable water all around it.

Description of isolated danger marks

Topmark:	2 black spheres, one above the other
Colour:	Black with one or more broad horizontal red bands
Shape:	Optional, but not conflicting with lateral marks; pillar or spar preferred
Light (when fitted):	
Colour:	White
Rhythm:	Group flashing (2)

Safe water marks

Definition of safe water marks

Safe water marks serve to indicate that there is navigable water all round the mark; these include centre line marks and mid-channel marks. Such a mark may also be used as an alternative to a cardinal or a lateral mark to indicate a landfall.

Description of safe water marks

Colour:	Red and white vertical stripes
Shape:	Spherical; pillar or spar with spherical top mark
Topmark (if any):	Single red sphere
Light (when fitted):	
Colour:	White
Rhythm:	Isophase, occulting, one long flash every 10s or Morse 'A'

Special marks

Definition of special marks

Marks not primarily intended to assist navigation but which indicate a special area or feature referred to in appropriate nautical documents, for example: Ocean Data Acquisition Systems (ODAS) marks.
Traffic separation marks where use of conventional channel marking may cause confusion.
Spoil ground marks.
Military exercise zone marks.
Cable or pipeline marks.
Recreation zone marks.

Description of special marks

Colour	Yellow
Shape:	Optional, but not conflicting with navigational marks
Topmark (if any):	Single yellow 'X' shape
Light (when fitted):	
Colour:	Yellow
Rhythm:	Any, other than those described in sections on cardinal, isolated danger and safe water marks.

Additional special marks

Special marks other than those listed and described above may be established by the responsible administration to meet exceptional circumstances. These additional marks shall not conflict with navigational marks and shall be promulgated in appropriate nautical documents and the International Association of Lighthouse Authorities notified as soon as practicable.

New dangers

Definition of new dangers

The term 'New Danger' is used to describe newly discovered hazards not yet indicated in nautical documents. 'New Dangers' include naturally occurring obstructions such as sandbanks or rocks or man made dangers such as wrecks.

Marking of new dangers

'New Dangers' shall be marked in accordance with these rules. If the appropriate Authority considers the danger to be especially grave at least one of the marks shall be duplicated as soon as practicable.

Any lighted mark used for this purpose shall have an appropriate cardinal or lateral VQ or Q light character.

Any duplicate mark shall be identical to its partner in all respects.

A 'new danger' may be marked by a racon, coded Morse 'D' showing a signal length of 1 nautical mile on the radar display.

The duplicate mark may be removed when the appropriate Authority is satisfied that information concerning the 'New Danger' has been sufficiently promulgated.

Characteristics and chart abbreviations

On Admiralty charts the position of a buoy, beacon, light vessel, *etc*, is shown by a small circle drawn at the centre of the base line. Large scale charts showing more detail of soundings and lights give buoys, beacons, *etc* drawn properly to shape; the colour or colours of the buoy are shown below the buoy's position – the small circle. The abbreviations for the colour of buoy's beacons are shown in capital letters.

The standard buoy shapes are cylindrical (can) ⌒, conical △, spherical ◠, pillar (including high focal plane) ⌿, and spar /, but variations may occur, for example: light-floats ⇌.

When buoys, *etc*, are lighted, this will be indicated by a magenta flare. The characteristics of the light, in abbreviated form, are printed to one side of the light as is its name or number, if any.

Light Characters

Metric and Fathoms Charts

CLASS OF LIGHT	International abbreviations	Older form (where different)	Illustration Period shown
Fixed (steady light)	F		
Occulting (total duration of light more than dark)			
Single-occulting	Oc	Occ	
Group-occulting e.g.	Oc(2)	Gp Occ(2)	
Composite group-occulting e.g.	Oc(2+3)	Gp Occ(2+3)	
Isophase (light and dark equal)	Iso		
Flashing (total duration of light less than dark)			
Single-flashing	Fl		
Long-flashing (flash 2s or longer)	LFl		
Group-flashing e.g.	Fl(3)	Gp Fl(3)	
Composite group-flashing e.g.	Fl(2+1)	Gp Fl(2+1)	
Quick (repetition rate of 50 to 79 - usually either 50 or 60 - flashes per minute)			
Continuous quick	Q	Qk Fl	
Group quick e.g.	Q(3)	Qk Fl(3)	
Interrupted quick	IQ	Int Qk Fl	
Very Quick (repetition rate of 80 to 159 - usually either 100 or 120 - flashes per min.)			
Continuous very quick	VQ	V Qk Fl	
Group very quick e.g.	VQ(3)	V Qk Fl(3)	
Interrupted very quick	IVQ	Int V Qk Fl	
Ultra Quick (repetition rate of 160 or more - usually 240 to 300 - flashes per min.)			
Continuous ultra quick	UQ		
Interrupted ultra quick	IUQ		
Morse Code e.g.	Mo(K)		
Fixed and Flashing	F Fl		
Alternating e.g.	Al.WR	Alt.WR	

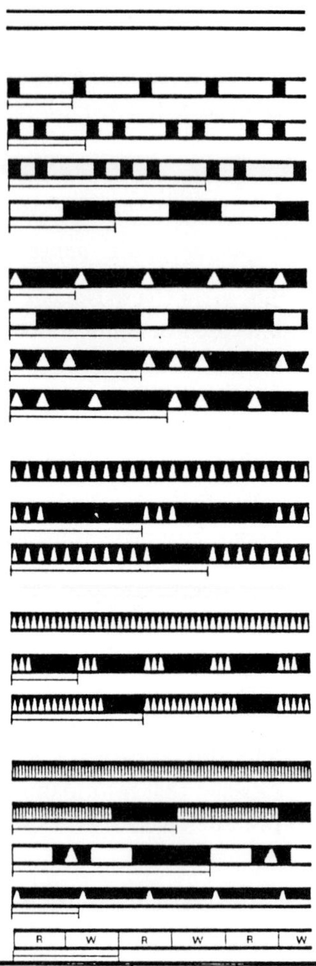

COLOUR	International abbreviations	Older form (where different)	RANGE in sea miles	International abbreviations	Older form
White	W (may be omitted)		Single range e.g.	15M	
Red	R				
Green	G		2 ranges e.g.	14/12M	14,12M
Yellow	Y				
Orange	Y	Or	3 or more ranges e.g.	22-18M	22,20,18M
Blue	Bu	Bl			
Violet	Vi				

ELEVATION is given in metres (m) or feet (ft)			PERIOD in seconds e.g.	5s	5sec

Fig 4.3 Light characters. Crown copyright. Reproduced from 'British Admiralty Charts with permission of the Controller of H.M. Stationary Office and of the Hydrographer of the Navy.'

In British waters, to avoid confusion with ships' navigation lights, fixed shore lights shall be shown in pairs disposed vertically. Alternatively a single red or green light may be used but it must be flashing or occulting.

In order to distinguish spar buoys from fixed beacons on Admiralty charts, spar buoy symbols will be sloped, in accordance with standard practice. Beacons will be shown in the vertical.

A lighthouse may show an alternating light as or it may show a sector light or lights. Sector lights usually show a red light over a danger area or approach, and a white light over a safe area or approach. Skippers should consult the chart and Admiralty Light List when approaching sector lights.

If lighthouses, lightships and buoys emit signals other than light signals, this will also be shown on charts using one of the following: diaphone, horn, whistle, bell, siren, racon, *etc.*

Lightships and lighthouses may indicate the height of the light above sea level, (MHWS) and the luminous range of the light.

The abbreviation, Mo, when shown, indicates a fog signal which consists of one or more characters in the morse code, *eg* Horn Mo (WA).

Large scale charts give more detail of lights and signals than a small scale chart and should always be used in conjunction with the volume of the Admiralty List of Lights.

5 Tides and tidal streams

Gravitational attractions of the sun and moon create fluctuations about the mean level of the sea and the resultant effects are known as tides.

The moon is approximately 400 times nearer to the earth than the sun and it exerts a force of attraction which is about 2¼ times greater than that of the sun.

The rise and fall of tides would closely follow the movement of the moon around the earth varied by the relative positions of moon, earth and sun, were the earth smooth with universal water depth, no wind, no land masses or islands. But in fact the way in which tides rise and fall in different seas and oceans varies considerably; the period of oscillation of the tide in one area differs from that in another and varies from about 6 to 24 hours. An oscillation of about 24 hours, called a diurnal (or daily) tide would result in one high water and one low water a day; one of about 12 hours, called a semi-diurnal (or half-daily) tide, would result in two high waters and two low waters a day; while a 6 hour period would result in four high and four low waters a day.

Springs and neaps

The combined tide raising forces of the moon and sun have their greatest effect when the sun and moon are in line with the earth, *ie* at new moon and full moon, and least effect when they are approximately at right angles to each other, *ie* at the first and last quarters of the moon, as shown in *Fig 5.1*.

These variations affect the difference in level between successive high and low waters (*ie* the range of tide). Shortly after full and new moon a locality will experience the highest high waters and lowest low waters of that lunar month and the tides at this period are called Spring Tides. Conversely around the first and last quarters of the moon will be experienced the lowest high waters and highest low waters of that lunar month, at which period the tides are called Neap Tides. Between these limits the heights of successive tides increase or diminish progressively.

The time interval between successive Spring and Neap Tides can in general be taken as being about 7½ days.

Interval between tides

The interval of time between one high water and the corresponding high water on the following day depends to a large extent on the period of a lunar day, which is about 24 hours 50 minutes. As a rough guide it can be taken that, with a tide whose period of oscillation is regular, high water will occur about 50 minutes later each day.

The interval of time between successive high and low waters will, of course, depend upon the period of oscillation. For a tide with a 12-hour period (such as that experienced around the British Isles) the interval between successive high waters would be about 12 hours 25 minutes, while the interval between one high water and the succeeding low water would be about 6 hours 12 minutes.

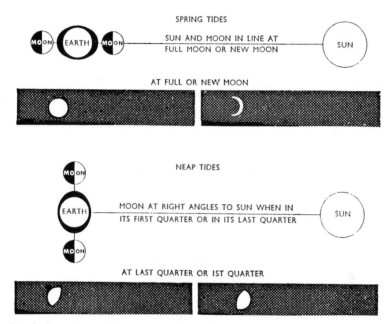

Fig 5.1 Spring and neap tides

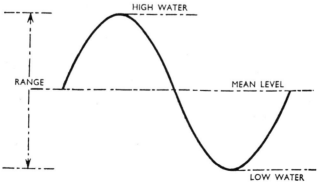

Fig 5.2 Graph of tidal oscillation

Tidal definitions

Knowledge of the following tidal definitions is required for chartwork and for use of the Tide Tables. They are illustrated in *Fig 5.3* with example of a particular tide.

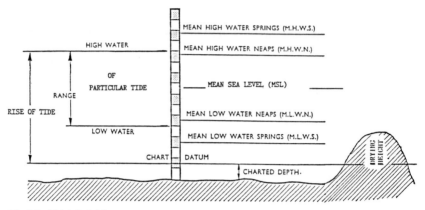

Fig 5.3 Tidal definitions

Mean sea level (MSL). The average level of the surface of the sea.

Tidal oscillation. A tidal wave represents one vertical oscillation about the mean level of the sea, and, as shown in *Fig 5.2*, it includes one high water and the succeeding low water.

High water (HW). The highest level reached by the sea during one tidal oscillation.

Low water (LW). The lowest level reached by the sea during one tidal oscillation.

Slack water. The period when a tide has ceased to rise or fall at high or low water, preceding a change of direction of the tide. The slack water period is usually of longer duration on neap tides and shorter on springs.

Chart datum. The plane from which the depths of all features, either permanently or nearly permanently covered by the sea, are measured. It is also the plane from which the heights of all features, periodically covered and uncovered by the sea, are measured. The former are charted depths and the latter, which are distinguished on the chart by an underline, are known as 'drying heights'. It is a fixed level below which all depths on a chart are given and above which all 'heights of the tide' are given in the tide tables.

Chart datum is at a level below which the tide seldom falls. On Admiralty charts this plane has been, in the past, usually near or below Mean Low Water Springs in the locality, but these datums are now adjusted to approximate to the level of the Lowest Astronomical Tide (LAT).

Lowest astronomical tide. (LAT) The lowest height which can be predicted to occur under average meteorological conditions and any combination of astronomical conditions. The level of LAT is reached only occasionally and not every year. Lower levels than this can occur with particular meteorological conditions such, for example, as storm surges. In the British Isles chart datum at all ports has been amended to approximate LAT. All metric charts of these waters are referred to this datum.

Soundings on a chart. Depths of water below the level of chart datum. They are given either in metres, fathoms or feet. Drying heights are always given in metres or feet, never in fathoms.

Height of tide. Is the vertical distance at any moment between the level of the sea and the chart datum. The height of tide does not include the amount of water which may be below chart datum, and consequently does not indicate a depth. In tide tables when the sea level falls below chart datum, the figure will be prefixed with a minus sign.

Depth of water. The sum of the charted sounding and the height of tide. In the case of a drying charted sounding, the depth of water is the difference between the height of tide and the drying height.

Range of tide. The difference in height between that obtaining at high water and that of the preceding or succeeding low water.

Rise of tide. Is the difference between high water and chart datum.

Tide tables

Many tide tables are produced giving times of high and low water at ports of interest to fishermen. But it is not always sufficient simply to know these times and heights, as a mariner often needs to know the height of the tide at intermediate times; or to find the time at which the tide may be expected to reach a certain height. In coastal navigation he will also need to be able to apply an approximate correction to soundings taken in relating these to depths indicated on a chart.

These requirements are met in Admiralty Tide Tables and other similar tables which give the times (GMT) and heights (in metres) of high and low water at standard and secondary ports, the latter by giving their tidal differences on the standard port. Admiralty Tide Tables also contain related tide curves for standard ports from which the actual rise of tide at intermediate times can readily be obtained by simple interpolation, both at standard and secondary ports.

The rate of rise or fall of a tide will not be uniform, and the extent of the rise or fall every hour will depend on the interval between low and high water. A rough guide to the rise or fall of a 6-hourly tide is given below, but it must be emphasised that this method of estimating the approximate height of the tide at any particular time is only a general guide. A 6-hour tide may be expected to rise or fall approximately:

1/12 of its range in the first hour
2/12 of its range in the second hour
3/12 of its range in the third hour
3/12 of its range in the fourth hour
2/12 of its range in the fifth hour
1/12 of its range in the sixth hour.

It will be seen that the maximum rate of rise or fall occurs at half-tide.

Convenient tables for determining the approximate height of the tide at times between high and low water, on this basis, are set out in Appendix 2 on page 289.

Applying height of tide to charted depth

Modern fishing vessels are usually fitted with sophisticated sounding equipment, which should be used as an aid to navigation as well as an aid to fish catching.

The depths shown on a chart are the minimum depths so that unless due allowance is made for the height of tide a dangerous error may arise when navigating in coastal waters. Let us suppose that a sounding from a ship shows a single or continuous depth of 10 metres. It is wrong for the navigator to point to a 10 metre mark or line on his chart and to say that he is in that position. His statement will be true only if it happens to be low water with the depth taken as being that of chart datum. This will rarely happen and at all other times a sounding of 10 metres will indicate that the ship is inside the 10 metre line. See *Fig 5.4.*

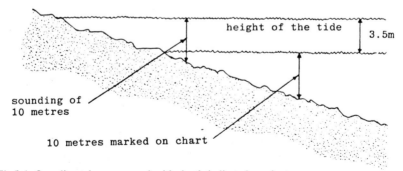

Fig 5.4 Sounding taken compared with depth indicated on chart

The problem of reducing the sounding taken to the depth on the chart is most readily solved for a busy skipper by reference to the table in Appendix 2.

The height of tide must then be subtracted from the sounding to arrive at the depth indicated on the chart for that position.

Mean tide level

An understanding of mean tide level is essential for the mariner who is to navigate in tidal waters with minimum under keel clearance. In ports where the tide is semi-diurnal and regular, the mean tide level is the same for all tides whether they be springs, neaps or intermediate tides. Local knowledge and examination of tide table predictions for a port will show whether the tide is regular or not. If there is a mean tide level of 4 metres on neaps there will be a mean tide level of 4 metres on springs or any other tide. The reason for this is that the tide falls below its mean level by the same amount as it rises above it. If we now look at our tidal diagram (*Fig 5.5*) with a mean tide level of 4 metres and a time scale of six hours, by drawing lines from each hour horizontally to the vertical scale we find that the regular tide follows a pattern. On the flood tide it will be seen that in the hour preceding the mean tide level the tide rises as much as it did in the first and second hours combined. In the hour following the mean tide level, the rise is roughly equal to the rise of tide during the two hours before high water.

From this it will be seen that the greatest rise of water taking place on the flood will be between four hours and two hours before high water. The velocity of the flood tide in estuaries will be greater during these two hours so that allowance for tidal effect in fog or when manoeuvring will have to be taken into account to a different degree at different times when on passage in the estuary.

If we assume that at a particular dock with a regular semi-diurnal tide the mean tide level at three hours before high water is 4 metres, then by watching the rise of water on the dock sill gauge move from 4 to 5 metres in one hour we know that the remainder of the flood will produce about 1 metre more during the last two hours, giving a total height of 6 metres as shown in *Fig 5.5*. It is therefore possible to estimate the rise of an actual tide.

To take a different example for a port with a mean tide level of 5.3 metres; we watch the rise of water for one hour from the time this mark is reached and this proves to be 1.5 metres, then we know roughly that high water would make 8.3 metres because the rise in the last two hours will approximately equal the rise in the preceding hour. By applying the drying heights in the dock channel against 8.3 metres the skipper can determine the depth of water in which he may sail at high water.

54

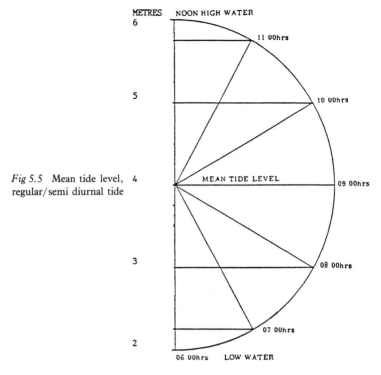

Fig 5.5 Mean tide level, regular/semi diurnal tide

METRES NOON HIGH WATER
6

5

4 MEAN TIDE LEVEL

3

2

11 00hrs

10 00hrs

09 00hrs

08 00hrs

07 00hrs

06 00hrs   LOW WATER

Local tides

The tides experienced in one locality may differ considerably in period from those experienced in another. Such differences may be apparent at places quite close to each other; for instance, although the tides around the British Isles are semi-diurnal in character, some places may have a double high water (Southampton) and some a double low water (Hook of Holland). In other parts of the world – some parts of the Pacific Ocean for example – a diurnal tide (*ie* a tide with only one high water and one low water a day) may be experienced, and many localities in the Mediterranean Sea experience little or no appreciable tide. Local knowledge of tides is therefore of great importance to the fisherman, and before visiting a strange port he should consult the Tide Tables for the behaviour of the tides in that locality.

Tidal streams

The influence of the sun and moon which create the tides as already described give rise to the horizontal movement of the sea from one place to another, where a tide is to be made. This movement is known as a tidal stream. These tidal streams, which are directly related to the time of high water, follow a fairly regular pattern and direction depending on the tide, *ie* a spring or a neap. They are indicated on a chart thus:

= Direction of ebb (with rate)
= Direction of flood (with rate)
= Direction of non-tidal current (with rate)

The Admiralty and various other nautical publications show tidal streams for different parts of the world. If we examine the Admiralty publications for the British Isles we find that there are 14 atlases which cover the United Kingdom. Each atlas shows the speed and direction of the tidal stream for each hour, beginning six hours before HW at Dover and finishing six hours after HW.

The directions of the tidal streams are shown by arrows; a lightly drawn short arrow indicates a weak stream, and an emphasised long black arrow indicates a strong stream. The mean strength of the stream is indicated between arrows, thus 20,35 indicates a 2.0 knot stream on neaps and a 3.5 knot stream on springs. The comma shows the approximate position of observation of streams.

Effect of weather on tides

Meteorological conditions will affect both time and height of tides. Gale force winds, depending on direction, may reduce or hold back a tide, particularly a neap tide. A gale or hurricane force wind blowing in the same general direction as a flood tide, will create a tidal surge of exceptional height and velocity if the tide is a spring. In estuaries and ports fed by rivers, the effect of heavy rains and melting snow creates a higher tide than predicted but because it runs towards the sea, it will reduce the speed of a flood tide and increase the speed of the ebb. The actual time of high water may well be earlier than predicted, with no slack water period at all and an early running ebb. Low barometric pressure also tends to raise the sea level and high barometric pressure to lower it.

Grounding

The extent of the damage a vessel will suffer through grounding in tidal waters depends upon her construction, the weather and the nature of the bottom. If the hull is intact, her prospects of floating off again given normal tides, and not being driven further on to the bank or shoal by wind and sea, would be as follows:

a vessel grounding on a rising tide will probably soon refloat

a vessel grounding on a falling tide, say, an hour after high water, will not float again until one hour before the next high water, *ie* about ten hours later in the British Isles.

a vessel grounding at the top of high water, say two days after springs, should not expect to float without assistance until about two days before the next springs, *ie* about ten days later.

6 Charts

Chart projections

The shortest distance between two points on the earth's surface is that followed by a great circle, which is curved. A great circle is a line circumscribing the earth, the plane of which passes through its centre. A portion of the arc of a great circle is, therefore, the nearest approach to a straight line which can be drawn on the earth's surface.

Mercator chart

To the navigator the most useful chart is one on which he can show the track of his ship by drawing a straight line between any two positions and so be able to measure the course from one to the other. The Mercator projection permits him to do this by showing longitudinal meridians as vertical lines running north and south and lines of latitude running at right angles to the meridians. Any line which crosses the meridians or parallels at the same angle is called a rhumb line and is not the shortest distance between any two points.

Scale on the Mercator chart

Since the equator (which is a rhumb line as well as a great circle) appears on the chart as a straight line of definite length and the meridians appear as straight lines perpendicular to it, the longitude scale is fixed by that length for all latitudes. To compensate for this distortion of the meridians of longitude the distance scale in any latitude is governed by multiplying the distance scale at the equator by the secant of that latitude. The effect is that the latitude scale on a Mercator chart and hence the scale of distance increases in direct proportion as the latitude increases. All distances must therefore be measured on the latitude scale to the sides of the chart. (When measuring distances between two points of differing latitudes, the navigator must use the scale at the mean latitude of the two places). The longitude scale is used only for taking off the longitude of a position, never for measuring a distance.

Gnomonic charts

Gnomonic projections are those upon which great circles appear as straight lines. Gnomonic charts are small scale ocean charts and are intended for use in plotting oceanic courses. When planning to follow an approximate great circle track sections of the gnomonic straight line course can be easily transferred to a Mercator chart by plotting the latitude and longitude of the ends of sections of convenient length.

The Mercator chart (*Fig 6.1*) shows the rhumb line or course from Cape Wrath, Scotland to Cape Farewell, Greenland as a straight line and seemingly the shortest distance. This is in fact incorrect. *Figure 6.2*, the gnomonic chart, shows the true distance of the rhumb line as being the longer.

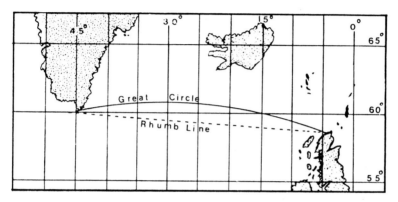

Fig 6.1 Mercator chart, showing rhumb line and great circle tracks.

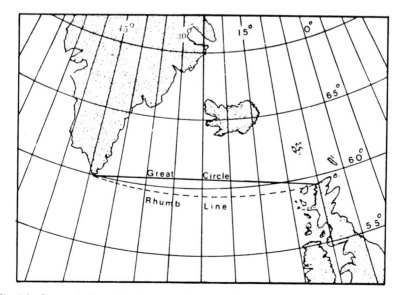

Fig. 6.2 Gnomonic chart, showing great circle and rhumb line

Consequently, a gnomonic chart should be referred to when navigating over long distances especially in high latitudes when east or west bound. A note of caution must be introduced here, however; it would be of little use saving time and fuel when UK bound from Newfoundland or Greenland if the great circle track found ice on the northern arc of the circle. Far better to pass well south of the ice before making a great circle sailing.

Latticed charts

Admirality nautical charts are available with coloured overprints showing the position-fixing lines of various electronic navigation systems. By far the most commonly used of these is the Decca Navigator, and latticed versions are available of most medium and small-scale charts of the whole of north-west Europe, including the Baltic Sea and South Africa. Rather more limited cover is available for the Persian Gulf, India and Pakistan, north-west Australia, and Japan. Omega Navigation System lattices for the 10.2 kHz basic frequency are available on small-scale charts covering most of the world's ocean areas. Charts with Loran-C overprints are being produced for those areas of the North Atlantic and the coasts of USA and Canada which are within the groundwave coverage of the system. In the case of Decca, Omega, and Loran-C, the lattices are overprinted on to the standard charts. They are therefore always fully corrected for ordinary navigational changes and may be used in a dual role, for navigation both with or without electronic aids. All such charts bear the letter 'L' in the lower right hand corner.

Small-scale chart cover of north-west Europe is also available with overprints for the Consol radio navigation system. This series is overprinted on bases which are not standard navigational charts and not kept corrected for notices to mariners. These charts are intended solely for use with the electronic aid and for ordinary navigation the standard chart of the appropriate scale must be consulted.

Loran charts are also published by the governments of Canada, Germany, and the USA. Loran is of particular importance to those fishing the NW Atlantic and Bear Island, Norway, and also in those areas out of Decca range.

Routeing charts give routeing information for each month of the year and include tracks, distances, meteorological and ice conditions.

Magnetic charts show variation and its changes for the various parts of the world.

Traffic separation schemes

These schemes which have been adopted by IMO (International Maritime Organisation) are of particular interest to fishermen. Traffic separation schemes are usually laid down in high density traffic areas and are clearly shown on charts.

A traffic separation scheme consists of two lanes in which traffic proceeds in opposite directions within separate lanes so as to avoid end-on situations. The two lanes will be divided centrally by a line or an area which is known as a *separation zone line*. Similarly the traffic lanes will have a *separation zone line* on the outer limits.

Charts to be carried

Following international regulations that 'all ships shall carry adequate and up-to-date charts, sailing directions, lists of lights, Notices to Mariners, tide tables and all other nautical publications necessary for the intended voyage' national administrations accordingly issue appropriate rules for fishing vessels.

British Admiralty charts normally cover most situations but there are countries whose local regulations require their national charts and publications to be carried when navigating in their national waters. There may also be occasions when a skipper may find it useful to supplement British Admiralty charts with foreign governmental charts or those produced by local authorities on a larger scale.

The catalogue of British Admiralty Charts and other Hydrographic Publications gives complete coverage of all its publications and comes out yearly. It is supplemented by a Home Waters catalogue covering North Western Europe.

Fishing charts, specially adapted to that purpose, should not be used for navigation. They are mainly concerned with marking the nature of the sea bed, are usually of small scale and may omit details of lights and shore marks.

The use of small scale charts should be restricted to passage planning; running off a long course, measuring overall distance and general observation of likely impediments.

In all coastal navigation the largest scale charts available should be used: these contain the most detailed and up-to-date information. Large scale charts are always corrected first, being under constant revision in the Hydrographic Department. As necessity arises, the Hydrographic Department issues new charts and new editions of existing charts. Notices of all such issues are published in Admiralty Notices to Mariners.

The age and state of a chart are indicated along its bottom margin and by comparing this information with the latest catalogue of Admiralty Charts the mariner can readily see whether it is the most up-to-date available.

The Revised Print (RP) Charts, introduced in 1972, were a new printing of an existing chart. They incorporated amendments not previously promulgated by Admiralty Notices to Mariners but did not constitute a new edition. Their purpose was to chart important items of navigational information which did not carry the same urgency as Notices to Mariners information but which should not be delayed until the next New Edition. Usually associated with high useage charts they are being discontinued to avoid possible confusion resulting from having differing charts of the same date. Revised Prints should not be confused with New Editions.

New editions and corrections

When a New Edition is produced the edition date changes, as shown on the bottom right hand corner of the chart.

Small corrections to charts are published in Admiralty Notices to Mariners to be made by hand, either at the chart agency or by the user. Such corrections are recorded at the bottom left hand corner of the chart; they are best readily checked by being entered in a log.

Admiralty Notices to Mariners are the only definitive sources of information on corrections to Admiralty Charts and publications.

Owners can arrange for Admiralty Notices to Mariners to be mailed weekly by Admiralty Chart agents to a vessel to ensure as regular supply as possible. Each envelope contains one Weekly Edition of Notices to Mariners, listing all the British Admiralty charts being corrected that week. Otherwise, the navigator can personally obtain British Admiralty Notices to Mariners free of charge from Admiralty Chart Agents and certain Mercantile Marine Offices and Customs Offices.

The importance of keeping charts up-to-date from corrections promulgated cannot be stressed too highly. If a vessel is involved in any accident concerning its navigation the actual chart in use will be studied in any action taken by Government Surveyors or by Marine Experts in pursuit of legal claims. There has been a recent instance of a vessel anchoring across and damaging an oil pipeline unaware of its existence from using an uncorrected chart.

7 Publications

For fishermen operating off unfamiliar coasts the following publications listed in the catalogue of Admiralty Charts and other Hydrographic Publications are of particular interest.

Pilots or sailing directions

Their purpose is to advise the mariner as a live pilot would do for navigation. They contain descriptions of the coast and off-lying features, notes on tidal streams and currents, directions for navigation in intricate waters, and other relevant information about the channels and harbours. In addition each book includes information about navigational hazards, buoyage systems used in the area covered, pilotage, regulations, general notes on the countries within the area, port facilities and a general summary of seasonal current, ice and climatic conditions. The series covers all the navigable waters of the world. The sailing directions are in process of being revised in style, avoiding unnecessary duplication of what may be found from a chart.

Lists of lights and fog signals

The several volumes, each conveniently grouping various areas of the world (so that fishermen would be unlikely to need more than one volume at a time), give a tabulation of all lighthouses and lights of navigational significance. Also listed are lightships, lit floating marks if over 8 metres in height and fog signals.

List of Radio Signals

This is in six volumes covering respectively:

Volume 1   Coast radio stations.
Volume 2   Radio Navigational aids (Beacons).
Volume 3   Radio weather services.
Volume 4   Meteorological observation stations.
Volume 5   Radio time signals. (Radio navigational warnings and position fixing systems).
Volume 6   Port operations, pilot services and traffic management (divided into two parts: Part I covers North West Europe and the Mediterranean. Part 2 covers the rest of the world).

8 Navigational instruments and equipment

The hand lead and line

This means of sounding, in common use before the advent of electronic navigational aids used for sounding and position fixing, is rarely used by the present-day mariner. Nevertheless, it still remains an important piece of equipment which is reliable in shoal waters. Mariners should be familiar with the markings on the leadline and should use the line for no other *purpose than sounding.*

The marking of a metric lead line

1, 11 and 21 metres	one strip of leather
2, 12 and 22 metres	two strips of leather
3, 13 and 23 metres	blue bunting
4, 14 and 24 metres	green and white bunting
5, 15 and 25 metres	white bunting
6, 16 and 26 metres	green bunting
7, 17 and 27 metres	red bunting
8, 18 and 28 metres	blue and white bunting
9, 19 and 29 metres	red and white bunting
10 metres	leather with a hole in it
20 metres	leather with a hole in it and 2 strips of leather
All 0.2 metre markings	a piece of mackerel line

The lead may be used to check the echo sounder for accuracy when the ship is stopped or when at anchor.

Arming the lead

The bottom of the lead has a cavity not unlike that of the bottom of a wine bottle which is filled with tallow to which sand, gravel and shells become embedded. These specimens with the corrected sounding help to fix the ship's position by reference to soundings on the chart and the abbreviations indicating the nature of the bottom.

Logs

Towed rotator logs

With the advent of impeller and pressure logs built into ships the towed log is not in common use nowadays. The use of Decca navigator and other position fixing systems has also caused a reduction in its use. This log is streamed from the stern of a ship and consists of a rotator, a fish, a long line, a governor wheel and a clock, with a shoe fitting into a plate screwed into the taffrail. (See *Fig 8.1*).

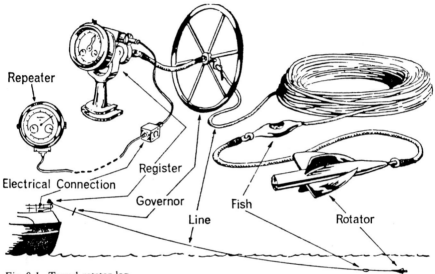

Fig. 8.1 Towed rotator log

Impeller and pressure logs

These logs are fitted into the bottom of a ship via a tube which projects through the hull into the sea. The impeller log, as its name implies, is rotated by passing through the sea when the ship is steaming and the revolutions are converted into distances and speed registers in the wheelhouse or chartroom.

The pressure log is similarly projected into the sea from the ship's bottom. The principle that pressure on the tube varies directly with the speed of the ship enables speed and distance to be recorded as it is with the impeller log. Since they project from the ship's bottom such logs are no longer widely used in fishing vessels.

Electro-magnetic log

A log now fitted flush with the ship's bottom from which a magnetic field is generated. The speed of the water passing through this field, when the ship is steaming, is measured and recorded as speed and distance.

Doppler speed log

This is a pulse system whose transducer is mounted flush with the bottom of the hull. Digital data outputs are available to provide data to radar, navigation systems and collision avoidance systems.

It is particularly accurate in shallow water.

NB: With the exception of the Doppler speed log it must be noted that it is speed and distance through the water which is being measured and *not* speed and distance over the ground. Tides and currents must be taken into account when using logs which are motivated by the action of water passing the ship.

The magnetic compass

The magnetic compass is made up of needles, a compass card (*Fig 8.2*) a pivot and a compass bowl. The needles which point to the magnetic north are fitted into the card so that the north end of the needles coincide with the north mark on the card. The float of the card, which is sapphire, sits on an iridium pivot in order to reduce friction to a minimum. The card, if not of the dry card type, is suspended in the bowl in either white spirit or alcohol and distilled water which will not freeze. The bowl is suspended on gymbals so that the card will remain horizontal regardless of the movement of the ship. The bowl is made of non-magnetic material, with a glass top, and the gymbals are carried in a binnacle, or fitted on some ships into a deckhead. If a binnacle is used it would be made of non-magnetic material, wood and brass or man-made glass resinated plastic.

The compass will, in theory, point to the magnetic north and this would be so if it was fitted on board a non-magnetic ship which was made and fitted out entirely in wood or plastic. But this is unlikely, so the compass on each individual ship has to be corrected for all the magnetic influences which will be exerted upon it by the inherent magnetism contained in the ship itself. These local influences are known as deviation. The compass adjuster, by using permanent magnets and soft iron, adjusts the compass so that it will point to the magnetic north. However, when a ship alters course and the ship's head points in a different direction the various magnetic influences alter the direction of the field relative to the compass which is still trying to point to the magnetic north. The compass adjuster will adjust the compass as near as possible for all the headings around a compass and if there are any small deviations of $1°$ or $2°$, he will leave a deviation card on board which shows the residual deviation on each point of the compass. The navigator will have to use this card, before setting a course to steer, or after taking a bearing which he intends to lay off on a chart.

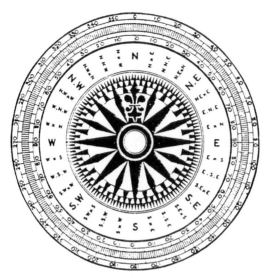

Fig. 8.2 Graduated magnetic compass card

We have determined that the compass, apart from the deviation, points to the magnetic north. But the north which we use for navigation, chartwork, *etc* is the true north or geographical north where all the longitudinal meridians converge. The magnetic north lies in a different position to that of the true north. The difference between magnetic north and true north is known as variation. It is found by looking at the chart in use where it is shown on the compass rose. Variation in any one place does not remain static but will increase or decrease from one year to another by a small part of a degree, expressed in minutes. Variation varies from place to place and will of course change as a ship moves along the surface of the earth. For the fisherman it will be of interest to note that on approaching Halifax, Nova Scotia from Newfoundland the variation alters by $10°$ in less than 500 miles but it varies in the English Channel by about $5°$ in 400 miles. Variation charts are available and should be consulted on these changes.

Rules for correction of courses and bearings

Before giving the helmsman a course to steer, the navigator will have to take the true course from his chart and correct it for variation and deviation. The combined effect of variation and deviation is known as compass error. Deviation and variation may be expressed as being either east or west. If they are of the same name, both east or both west, they are added together to make the error. If not, the difference becomes the error and takes the name of the larger.

Example

Variation	10°E	Variation	10°W	Variation	10°E
Deviation	2°E	Deviation	2°E	Deviation	2°W
Error	12°E	Error	8°W	Error	8°E

Before steering a course by compass the navigator must change the true course to a compass course by applying the error. When a compass bearing has been taken it must have the compass error applied to it so as to convert it to a true bearing for use on the chart. The compass card and true compass rose on a chart, the two points of interest, are marked from north which is zero to 360°, in a clockwise direction. Apply the error as follows:

From compass to true
Westerly error minus
Easterly error plus

From true to compass
Westerly error plus
Easterly error minus

If, however, the navigator wishes to use the magnetic compass on a chart, which of course differs from the true compass rose by the local variation, then only the deviation should be applied as follows:

From compass to magnetic
Westerly deviation minus
Easterly deviation plus

From magnetic to compass
Westerly deviation plus
Easterly deviation minus

Always remember that when applying deviation to a course or bearing the deviation used should be that which applies for the deviation of the ship's head when the bearing was taken, or for the course which is to be steered.

Marks on a compass card

The 32 points of a compass have for many years been used as the principal marks. Each point of $11\frac{1}{4}°(\times 32)$ made up the 360°. By looking at the figure it will be seen how a particular direction could be given in points, as follows. (See *Fig 8.2*).

North	*East*	*South*	*West*
N × E	E × S	S × W	W × N
NNE	ESE	SSW	WNW
NE × N	SE × E	SW × S	NW × W
NE	SE	SW	NW
NE × E	SE × S	SW × W	NW × N

ENE	SSE	WSW	NNW
E × N	S × E	W × S	N × W
East	*South*	*West*	*North*

Half and quarter points between the whole points can be used but references to courses and bearings are only accurate to about 2½°. This type of compass work and reference prevailed for many years and has been handed down to us by our forebears. It is simpler and more accurate to use the 360° compass when giving courses and bearings, particularly when passing information by radio. Always make sure that the bearing or course when used either by word of mouth or written is given its proper annotation (T) true, or (M) magnetic.

Gyro compass

Gyro compasses are mechanical and as a result are not affected by either the magnetism of the ship or that of the earth and are therefore not subject to variation or deviation. The gyro compass readings are always related to true north and the card is marked in the three figure notation only, ie 0° – 360°, in a clockwise direction. Thus NE by a gyro when written would be 045°T, when spoken, 'zero four five degrees true'.

The directive force of a gyroscope is provided by the rotation of the earth. It works on the principle that the axis of a perfectly balanced wheel suspended in gymbals, in such a manner that it is free to tilt or turn, will continue to point in a fixed direction as long as the wheel is spinning at a very high speed. By rotating the wheel mechanically and at high speeds, the axis of the wheel will point to the true or geographical north pole. A compass card is fitted to the mounting of the gyroscope, so that the north and south shown on the card, is parallel to the line of the wheel's axis. Whatever way the ship turns, the north on the card will point to the true north pole.

The gyroscope is, whenever possible, positioned on the centre line of the ship at the roll and pitch centre, so that the movement of the vessel affects it as little as possible. The gyroscope is electrically powered and should be run up for a period of five hours before use, so that it will be settled and reliable. From the gyroscope, repeater compass cards may be run off by electrical transmission to other parts of the ship – usually the wheelhouse or bridge wings for steering and the taking of bearings. Repeaters for use by the helmsman may take the shape of a roller tape, showing 30° – 40° of tape through the small window and will be lit for the helmsman to steer by. The lubber line is a marker or pointer which is in line with the fore and aft line of the direction of the ship's head.

Repeater compasses which are for use in taking bearings will be slung in gymbals and will be provided with either an azimuth mirror or bar sight. The azimuth mirror is a prism fitted on top of a compass bowl so that it can be turned through 360° around the compass card. A shore bearing or heavenly body by means of the reflective prism can be brought down on the edge of the compass card so that an accurate bearing between the observer's eye, the centre of the compass, and the edge of the compass card can be taken.

Gyro error

As long as the gyroscope is in good order and there is a steady input of electrical power, the repeaters will give satisfactory service, but the navigator must be aware of a small error which may have to be applied to the gyro. This error will be fairly constant and is due to course, speed and latitude. The axis of the gyroscope takes up a north-south position and it is, of course, at right angles to the earth's direction of rotation. When a ship is steaming east or west there is no steaming error, but when a ship steams due north, there is a slight deflection of the axis and the compass north, to the west of the true meridian. When steaming south there is a similar deflection of the compass north, to the east of the true meridian. The deflection is greater in high latitudes. The error may be found in tables.

Gyro error can be determined by observation of a transit. Also by observation of the sun's true bearing on rising or setting and comparing with amplitudes for that day from a nautical almanac.

The gyro-compass is marked numerically from $0° - 360°$ clockwise, so that if it is found that there is an error and the gyro-compass indicates a direction which is numerically greater than the true course or bearing then it is called *high*. Conversely a numerically smaller reading of the gyro-compass than the true is known as *low*.

The simple rules for applying gyro error are as follows:-

When converting direction from *compass to true, subtract* the gyro error if *high* and *add* the gyro error if *low*.

When converting from *true* direction to *compass* direction, *add* the gyro error if *high* and *subtract* gyro error if *low*.

Example:-
— A vessel is steering $045°$ by gyro compass and there is a gyro error of $2°$ high. What is the ship's true course?

Ship's head compass	$045°$	
Gyro error	$2°$	High (subtract) compass to true
Ship's true course	$043°$	

— A vessel has to make a true course of $130°$ but there is a gyro error of $2°$ high. What course must she steer on the gyro compass?

True course to make	$130°$	
Gyro error	$2°$	High (add) true to compass
Compass course to steer	$132°$	

— Flamboro Head bears $270°$ by gyro compass, gyro error $2°$ high. What is the true bearing?

Gyro compass bearing $270°$

Gyro error $2°$ High (subtract) compass to true

True bearing $268°$

Aneroid barometer

The aneroid barometer (*Fig 8.3*) is a dry barometer which contains a metallic chamber, partially exhausted of air and hermetically sealed. Variations in pressure of the earth's atmosphere allows the volume of air within the box to expand or contract. This change in volume is passed by means of levers to the hand on the face of the barometer so that an increase of pressure makes the hand rise and a decrease causes it to fall. There is no correction to be applied to an aneroid barometer except for the height above sea level and index error. An aneroid barometer should be checked fairly regularly against the corrected reading of a mercury barometer.

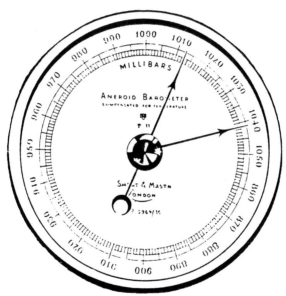

Fig 8.3 The aneroid barometer

Mercury barometer

The mercury barometer is a more reliable barometer than the aneroid, but more expensive and more difficult to fit into the confined space of a trawler's wheelhouse. The mercury barometer consists of a long metal case which holds the tube and cistern containing the mercury. The barometer is fitted to a bulkhead on a base plate and is slung in gymbals so that it remains upright, regardless of the movement of the ship. The mercury within the tube has above it a space which is exhausted of gases and the atmospheric pressure on the mercury in the cistern at the bottom of the tube balances the column of mercury. The fluctuations in pressure are shown as movements in the length of the column of mercury, against which a scale in millibars has been accurately set. The scale is graduated in millibars and marked at intervals of 10 MB. The scale may be read to an accuracy of 0.1 MB by means of a vernier scale.

When reading the mercury barometer the observer's eye should be exactly on a level with the upper edge of the mercury column so as to eliminate parallax. The surface of the mercury will be convex in shape and the reading should be taken by turning the vernier screw so that the lower edge of the vernier just touches the uppermost part of the domed surface of the mercury.

Corrections to the mercury barometer have to be made for height, latitude and temperature. There may well be an index error on the instrument which will be given on the maker's certificate issued with the barometer, as will be the standard temperature for the barometer. The principles of correction are as follows:

(1) The mercury in the barometer falls 1 MB for every 30 ft of height. Always add.
(2) The correction for latitude is made for the changing effect of gravity, greater at the pole than the equator. Consequently, the barometer reads lower than standard when the latitude is more than $45°$ when the correction is plus. Latitude less than $45°$ correction is minus.
(3) The mercury rises 1 MB for every $6°$ C so that the difference in the temperature taken from the thermometer attached to the barometer and the standard temperature on the certificate for the barometer must be applied. If the temperature is greater than the standard, apply minus correction, if temperature is less, apply plus correction.

Correction tables may be found in all good nautical almanacs.

Barograph

The barograph is a type of aneroid barometer which transmits expansion and contraction in the box with a partial vacuum to a roll of paper via a lever and pen (as opposed to a dial). The roll of paper is turned by clockwork so that the roll completes one revolution in seven days enabling a complete pressure record to be kept. The vertical lines on the barograph roll represent time and the horizontal lines represent pressure.

Thermometers

There are two main types of thermometer used by seamen, Fahrenheit (F) and Centigrade or Celcius (C). A third Absolute or Kelvin (A) is less commonly used. The scales used for these are:

	$F°$	$C°$	$A°$
Boiling point	212°	100°	373°
Freezing point	32°	0°	273°
Scale	180°	100°	100°

From the above scales it can be seen that C and A have the same scale and that 1° of Centigrade is equal to 1° of Absolute. The latter has, however, a range from 0° to 373° so that all temperatures below freezing point are positive. In the case of Centigrade, however, any temperature below freezing will be shown as negative.

When converting from F° to Centigrade, deduct 32° and multiply by 5/9. When converting to F° from Centigrade, multiply by 9/5 and add 32°. When converting from C° to A° add 273°, from A° to C° subtract 273°.

Temperature Conversion

$$C = 5/9(F - 32)\dots\dots\dots \qquad F = 9/5C + 32\dots\dots\dots$$

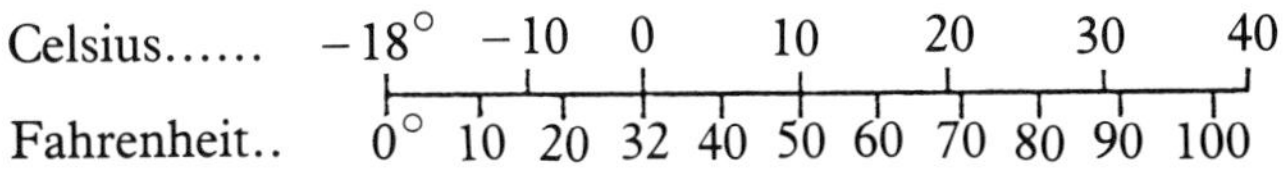

Celsius...... −18° −10 0   10   20   30   40

Fahrenheit.. 0° 10 20 32 40 50 60 70 80 90 100

Fig 8.4 Temperature conversion

The maximum thermometer

It is designed to record the highest temperature during a given period. The tube of the thermometer is reduced in bore close to the bulb. The thermometer in the horizontal position, with the temperature rising, allows the mercury to expand and force its way past the constriction so that when there is subsequently a fall in temperature, the mercury contracts below the constriction, leaving the mercury in the upper tube to remain as a record of the highest temperature reached.

Minimum thermometer

Designed to record the lowest temperature during a given period. Spirit is normally used and there is a small index, shaped so that when the temperature falls the index sinks down the stem towards the bulb. When subsequently the temperature rises the spirit passes the index and leaves it to register the lowest temperature.

Hygrometer

Consists of two ordinary thermometers, placed side by side, one of which is known as the wet bulb. The dry bulb records the temperature of the surrounding air. The wet bulb is wrapped in muslin with a wick leading to a canister of water which keeps the bulb damp. When the air is dry and has little water content, evaporation takes place on the surface of the wet bulb and because heat is extracted, the temperature will fall. The difference between readings on the two thermometers is of significant value.

A consistent difference of reading of say $10°$ is a sign of good weather, but any significant reduction in the difference may indicate the approach of a depression. At the other extreme, we may have a situation whereby the temperatures are similar. This happens when the air is so saturated that the evaporation cannot take place and mist/fog is forming.

The sextant

The sextant is an optical instrument used for measuring the angles subtended between any two points. Angles up to $120°$ may be measured even though the sextant's arc is only $60°$ or a sixth of a circle, hence the name sextant. (See Fig 8.5) Any type of angle may be measured but the navigator is mainly concerned with those on either the horizontal plane or the vertical plane. When out of sight of land the vertical angle between a heavenly body and the horizon is a basic necessity for working out an astronomical position line.

In coastal navigation the sextant can be used to calculate a distance off a known point of land or lighthouse by measuring its vertical sextant angle, *ie* the angle between the top of the object, the observer's eye and the waterline base of the object. The navigator must know the height of the shore object from the summit to the base and the vertical angle subtended at the observer and by use of 'Distance Off' tables and a bearing can obtain a fix.

The sextant can also be used to obtain a fix in coastal navigation by measuring two horizontal angles between three chosen points and plotting them by use of a station pointer or drawing on a tracing. This method is especially useful when, for any reason, it is not possible to take compass bearings. It also avoids any errors that might arise over compass errors.

By looking at *Fig 8.5*, it can be seen that the index arm is pivoted at the top of the sextant where an index mirror is fitted. The bottom of the index arm is able to slide along the arc which is graduated in degrees with a more accurate reading to a minute or part of a minute obtained from the micrometer. The index arm is clamped to the arc, but by a simple finger press may be released for movement.

In line with the telescopic sight, fitted to the frame of the sextant is a fixed glass, half mirror, half plain, which is called the horizon glass.

When the index arm is at zero on the arc the index mirror should reflect an image alongside the true image seen on the horizon glass. Shades are fitted at both the index and horizon glasses for use in bright weather and to darken the glare of the sun when it is being used.

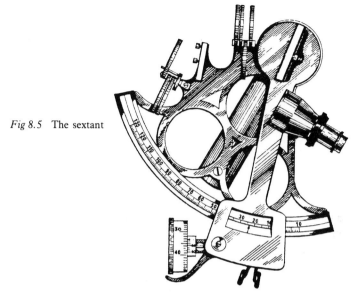

Fig 8.5 The sextant

Adjustable sextant errors

Perpendicularity

The index glass must be perpendicular to the plane of the instrument. Set the index bar near the middle of the arc. Hold the instrument horizontally, and look obliquely into the index glass. The reflected image of the arc should now be seen in line with the arc itself. Should the two not appear in line, they can be made to do so by rotating a small screw in the centre of the frame of the index glass.

This error must always be corrected first.

Side error

The horizon glass must be perpendicular to the plane of the instrument. Place the index at zero, hold the instrument with its plane nearly horizontal and look through the horizon mirror at the horizon. If the true and reflected horizons appear in one line the mirror is perpendicular, but if not they are brought into line by gently moving a screw at the back of the horizon mirror.

Index error

The horizon glass should be parallel to the index glass when the index is at zero. Place the index at zero, hold the instrument vertically and look through the horizon mirror at the horizon. If the true and reflected horizons are in one line the horizon mirror is parallel to the index mirror, but, if not, they must be brought into line by gently moving a screw at the back of the horizon mirror.

If this third adjustment to be made is very small it may be preferred not to make the adjustment but to find the index error. This can be found in several ways, the easiest being:

— By observing a star. Set the index bar a few minutes one side or other of zero; then with the sextant held vertically observe a star and bring the two images together so that one is indistinguishable from the other. The reading is the index error, minus (−) if 'on' the arc, and plus (+) if 'off' the arc.
— By observing the horizon. (This method is less accurate from a trawler in any kind of rough weather.) Set the arm at zero on the arc. Hold the sextant vertically and look through the horizon mirror to see the true horizon and reflected horizon. Move the micrometer until the horizons form an unbroken line. The reading is the index error.
    When the reading is on the arc, the index error is subtracted from future readings.
    When the reading is off the arc, the index error is added to future readings.
    NB: It is not a good policy to correct small index errors (under 3') since there is always a danger of straining the horizon glass and making the adjusting screw slack. A small index error should be allowed for in the angle measured.

Once his sextant is properly set up, a mariner should normally be concerned with only one of the adjustable errors that need regular checking: that is Index Error.

Use of sextant

To observe a vertical angle, clamp the index at zero on the arc and look at the object through the telescope and horizon glass. The reflected image will appear in the horizon mirror. To bring the object down to the horizon unclamp the index arm: move the sextant downwards, while moving the index bar away, in such a manner that the reflected image remains in the mirror, and until the horizon or lower object has been reached. Release the clamp with the left hand and make the final adjustment with the micrometer wheel. Remember that the left hand keeps the index mirror at the top of the sextant arm pointing at the object and that the right arm moves the sextant, its arc and reflection down until the reflected object has reached and is coincident with the second object. Several practice shots will soon make the student proficient.

To observe an horizontal angle hold the sextant in the right hand, arc up and with the index arm at zero. Look through the telescope and horizon glass at the right hand object which will be seen coincident in the glass and mirror. With the sextant held in the horizontal plane by the right hand, unclamp the index arm. Keep the index mirror pointing at the object with the left hand and swing the sextant slowly left with the right hand until the second object is reached and is seen by the eye in the horizon glass. Release the clamp and adjust the micrometer with the left hand until the objects are finally coincident in the horizon mirror.

Reading the sextant

The arc of the sextant is marked in degrees only, from $0°$ to $130°$ on the arc, and from $0°$ to $5°$ off the arc. The scale is numbered at every $10°$ of arc and each $10°$ section is marked in single degrees with a longer mark on each intermediate $5°$. The micrometer indicates the number of minutes on the wheel of the drum (see *Fig 8.5*) indicated by an arrow. To find the number of seconds we must now look to the vernier scale to the right of the micrometer wheel. There we will see one minute broken down into fractions, *ie* six graduations, each of which equals ten seconds. The arrow is zero, and six lines indicate the seconds. Where any one of these fixed lines comes into true conjunction with a line on the micrometer wheel, then that is the number of seconds to be used with the previously noted degrees and minutes. (On the illustrated sextant the reading is $23°$-$33'$-$20''$.) Some sextants have the minutes shown on the vernier as being in five parts, *ie* twelve second intervals. The student must understand the sextant and its readings, and only by practice will confidence in its use be attained.

The station pointer

This instrument made of brass or plastic consists of a circular frame which is marked off in degrees. From the centre of the circle running outward past the circular frame there are three legs. The middle of these three legs is fixed to the frame at zero degrees. The outside legs are movable at the centre circle pivot and are clamped on to the outer circle. They may be swung around the circle by releasing the pressure on the clamps by use of the knurled finger screws. By finding two angles subtended by three shore objects, the legs may be set to subtend these angles. By placing the station pointer on the chart so that the bevelled edges of the legs run through the positions of the objects, the position of the ship is fixed as being the centre of the circle. (*Fig 8.6*).

The angles may be determined either by sextant or by taking three bearings on the compass. If no station pointer is available tracing paper may be used.

76

Mark a spot on the tracing paper and from it draw a base line. Use this base line as the middle of the three bearings or angles taken by sextant. Measure the angles subtended by the other two shore objects from this base line, and use the tracing paper over the chart as the station pointer would be used. *Note:* when bearings are used it does not matter whether they be true, magnetic or compass as long as they are all the same and taken simultaneously. It is the difference between the bearings which gives the required angles.

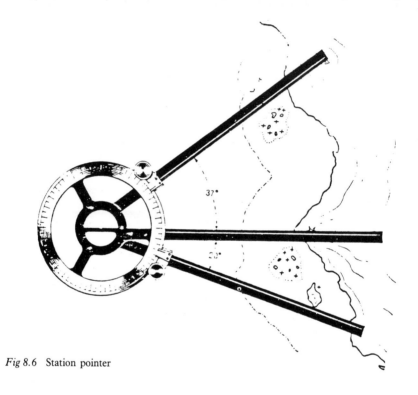

Fig 8.6 Station pointer

Automatic steering gear

The use of automatic steering gear has become widespread in both merchant and fishing fleets. It is said that it is more economical on fuel because it steers a ship on a straighter course and saves time.

There are many different types of automatic steering gear, too numerous to mention in this book, but they work on similar principles. One of the original types still in use consists of a brass ring fitted on to the top of the compass bowl. This ring is marked from $0° - 360°$ so that the compass card direction and that on the steering ring are the same. When a new course has to be steered the ring is moved so that the new course coincides with the lubber line. The automatic steering gear alters the course until compass card heading and that on top of the bowl coincide.

A second type is that of a black box, fitted to a bulkhead, marked with an illuminated ring from $0° - 360°$. From the middle there is an arrow pointing to the degrees on the ring from the middle of the circle, which indicates the course being steered. The arrowed indicator is fitted to a centre knob which can be used to select a fresh course by simply turning the knob and arrow to the new course required. The compass card heading should always match the black box heading once the course has been set. In both the above types of automatic steering there is usually a limit to the amount by which the navigator can alter course. Because the system is controlled by sensors, which put helm on according to the amount of the alteration, the alteration at any one time is usually restricted to a maximum of $30°$. When the ship's head has moved towards this maximum alteration then the navigator may apply a second maximum and so on. The danger in altering course by let us say $90°$ is that the sensors would seek the new course by putting the steering gear hard over and would continue to try and put more helm on. The ship would take on a violent swing and the sensors may well lose control. It is always advisable to make successive alterations limited to less than $30°$ with this type of steering. Manufacturers' instructions should be read and fully understood. Combined with automatic steering there is usually a steering tiller to which one may switch over. There may be more than one tiller, possibly one in the wheelhouse and one on each bridge wing.

The third type of automatic steering is that whereby the ship may be steered by the tiller mode as described above. The tiller is usually a small polished round bar about six inches in length with a knob at the end. It is usually horizontal and is housed in a box fitted on to the fore part of the wheelhouse in such a position that the helmsman can see not only the fore part of the ship but the compass and helm indicator. By moving the tiller to port or starboard the rudder is moved in that direction. When released the tiller handle comes back to the midship position. To take off the helm which has been put on, the tiller bar has to be moved across to the other side. On this third type of steering, course is altered by using the tiller until the ship is steady on the new course. The change-over to automatic steering may then be accomplished by pushing the hand tiller down into a groove marked automatic. The compass course is then maintained by the automatic sensors.

The advantage of automatic steering gear is that it allows a vessel to maintain a good straight course and a check on the course can quickly be made simply by looking at the compass occasionally. Unfortunately, this very useful equipment has led to some lack of bridge discipline in that a bridge may not be manned for short, and on occasions, long periods. The need to maintain a continuous and efficient lookout is most important. Because an officer of the watch knows that a good course is being steered by automatic means is no excuse for him to neglect to keep a proper lookout. There have been many cases whereby ships have been proceeding on automatic steering when they have run into another vessel or gone aground, because the officer of the watch felt it safe to go into the chartroom for a lengthy period, or the lookout felt it safe to leave the bridge for some reason. The automatic steering may steer very efficiently, but it cannot keep a lookout.

When in congested waters, estuaries and rivers, the hand mode of steering should *always* be used. In such areas any malfunction of the automatic gear can lead to collision or grounding almost immediately.

Pelorus

This instrument is constructed to be used in conjunction with a compass under circumstances whereby the subject of a bearing cannot be seen from the compass position.

In many trawlers the compass or gyro repeater may be in the wheelhouse or sometimes in the deckhead of the wheelhouse, where it is either difficult or impossible to take an accurate bearing of a landmark, seamark, *etc*.

The pelorus, a circular brass plate marked $0° - 360°$, is mounted on a pedestal, which fits into a shoe mounting available on each bridge wing. From the centre of the pelorus compass rose there will be a bar sight by which the observer may take a visual bearing. The bar sight may be moved in the horizontal plane around the compass rose. With zero on the pelorus set in the fore and aft line of the ship, a bearing which is relative to the ship's heading may now be taken. Care is needed to ensure that the bearing is taken when the ship is exactly on course or that the ship's heading is noted exactly at the time the bearing is taken. Station pointer bearings may be taken by pelorus without recourse to the compass if two or more landmarks are visible.

9 Coastal navigation

Navigation is the science and art of conducting a ship safely from one place to another. Generally we may consider modern navigation as falling into three categories – coastal, electronic and astronomical. This is a convenient method for separating the three types of navigation for discussion or reading, but it is important to remember that the different types of navigation may be used together. We may have, for example, a single bearing taken by compass of a distant point of land, which when used with a radio bearing or an astronomical position line will give a good positional fix.

When it is necessary to go from place A to place B, the *course to be made good* is the line joining the two places. This will not necessarily be the *course to steer* which will have to be found by allowing for the effects of tide, current, wind and compass error.

A single bearing, an astronomical intercept, a single radio bearing, or a radar bearing will always give the navigator a line which can be put on the chart, and on which the navigator knows he stands. This single line is known as a position line and if the navigator can obtain a second position line, where the two lines cross will give a position. Coastal navigation largely consists of the use of position lines, referred to as bearings. These bearings may be put down on a chart to provide a position or fix. To make sure that a correct course is being made good, intermediate position lines must be taken whenever possible so that a check is made on the true direction. If we suppose that a prominent point of land lies in a direction by compass 180° from the observer, then the *bearing* of the land is 180°. If we reverse the bearing and draw a line on the chart 360° from the point of land we then have a position line on which the observer stands. If we are able to take a second bearing simultaneously, let us suppose of yet another point of land bearing 090°, reverse it to 270° and lay it off from the point of land, we have a second position line on which the observer stands. Where the first and second line cross each other on the chart is the observer's position, known as a fix. Two bearings may not always be necessary. By looking at the chart two convenient and conspicuous objects may be marked; a chimney or church may be seen and if the observer waits until either the church or chimney is in line with let us say a small island, then a line drawn through both will provide a transit bearing. By taking a single compass bearing and laying it off on the chart as a position line, where it crosses the transit bearing will be the fix.

The best positions are always obtained from bearings which cross at a large angle, the ideal being 90°. It will be obvious that two shore objects which, when looked at by an observer, are so close together that the angle subtended at the observer's eye is about 20°, will not be very reliable because 20° is the angles of convergence and they run together slowly. Therefore an error of 1° in the bearing will separate the lines by some considerable distance. Do not if possible use angles of less than 40°.

The primary method of fixing should, whenever possible, be by means of visual bearings. In trawlers this may entail the use of the pelorus on which a relative bearing is taken and applied to the ship's head.

A quickly taken range and bearing obtained from a radar picture of a mark not positively identified is no substitute for the fix by compass.

Fix by cross bearings

Fig 9.1 illustrates an observer who took two simultaneous bearings of points A and B, bearing 317° and 027° respectively. When drawn on the chart the position lines cut clearly with an angular separation of 70°.

On the same figure the course to be made good (090°T) is already on the chart but the observed position is at C. It is clear that relative to the land, the true position by observation is on the landward side of the course line and that the ship may have to be pulled out or, if no danger exists, a new course line will have to be drawn from C with an allowance for the northward set which has been experienced.

Fix by vertical angles and a bearing

In coastal navigation a position line based on range will not be straight because it forms part of a circle. If we look at *Fig 9.2* we will see a lighthouse. The height of the focal plane of the lighthouse is known to be 40 metres above sea level (MHWS) and is taken from the charts. Let us suppose that we use a sextant to find the vertical angle between the top of the lighthouse, the ship, and the base of the lighthouse. If we find that the vertical angle is 0°31' then with the height of the lighthouse we may now find from published 'distance-off' tables (see Table at Appendix 3) that the ship is 2.4 nautical miles distant from the lighthouse. *Fig 9.2* illustrates the procedure, the top drawing shows in elevation the sextant angle and the bottom drawing the plan view and the position line as an arc of a circle drawn with the point of the compasses at the lighthouse as the centre of the circle. A bearing taken together with the sextant angle will provide a fix.

If the observer needs to pass the lighthouse at a given distance off for any reason, let us say five miles away from a lighthouse of 46 metres in height, look up the height and distance in the 'distance-off' tables to find the angle to set on the sextant $(0°17')$. By looking through the telescope of the sextant when abeam of the lighthouse, the reflected image should be on the waterline at the base of the lighthouse. If it is, the distance off is five miles. If the reflected image is below the waterline the distance off is more than five miles.

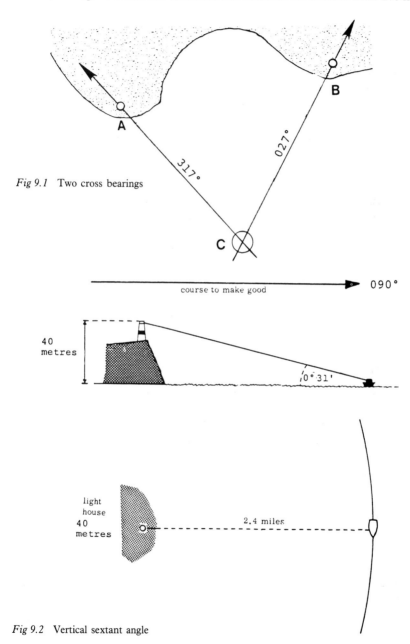

Fig 9.1 Two cross bearings

A   317°   027°   B

C

course to make good   090°

40 metres   0°31'

light house 40 metres   2.4 miles

Fig 9.2 Vertical sextant angle

If the reflected image is above the waterline, the distance off is less than five miles and it will be necessary to pull out by altering course.

(*NB:* For accurate fixing allowance must be made for the amount the tide is below MHWS).

Fix by station pointer and horizontal angles

By taking simultaneous compass bearings well spread out on one side of the ship a good fix will be obtained. As described in *Chapter 8, Station pointer,* three bearings of suitably spaced points of land will provide two horizontal angles subtended from the middle bearing. By examining the bearings taken and illustrated in *Fig 9.3*, the station pointer can be set with the angles 50° – 60° from the middle leg and by placing it on the chart so that the three legs pass over the three shore marks the position is fixed at the point of intersection.

By using a horizontal angle set on a sextant, an off-shore danger can be cleared as illustrated in *Fig 9.4*. Place the point of the compass on the off-shore danger D, draw the circle with the safe distance off as a radius. Now draw the course to be made good so that it just touches this small circle to seaward. This point should now be joined by straight lines to the two fixed identifiable objects ashore, A and B, one on each side of the danger. The angle of 53° subtended between A and B at position C, is the largest that can be allowed to occur between the fixed objects ashore. When set on the sextant it will indicate to the navigator whether he is passing clear of the danger or not.

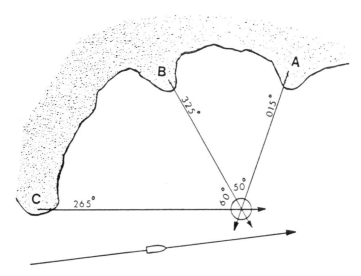

Fig 9.3 Fix by horizontal sextant angles and station pointer

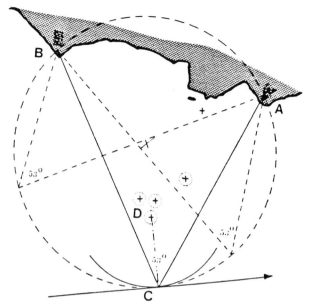

Fig 9.4 Horizontal safety angle

If, through the sextant's telescope, the right hand object A is reflected to the left of B which is seen directly, the ship is outside the danger circle. If however the reflected object appears to the right of the directly visual object B, the ship is inside the danger area and should be pulled out, in this case to starboard, until the objects appear together on the horizon glass of the sextant. (See *Chapter 8, Use of sextant*)

Use of radar

Today not many fishing vessels have sextants aboard and skippers therefore commonly fix the vessel's position by radar. Provided the shore objects selected are well defined to produce good radar responses the technique of fixing by radar range and visual bearing is to be preferred, though both radar range and bearing may often have to be relied upon: or simply two radar ranges as in *Fig 9.7*. When wishing to maintain a course at a clearing distance from dangers, the system used is that of parallel index, whereby the radar echo of the selected object on which to run is kept moving on a line which is parallel to one's track at the required distance.

Running fixes

A running fix is used to find a position when only one identifiable object ashore can be used. It is the least reliable type of coastal fix because its accuracy depends on the estimated run of the ship between two bearings of the same object. However, the running fix demonstrates that once having obtained a single position line which the navigator knows the ship is on, it can be transferred to cut with a second position line. The fix, however, is only as good as the estimated course and distance made good. To obtain a good running fix, the navigator must have a reasonably angled bearing with a fair idea of the course and distance run.

In *Fig 9.5* at 1050 a ship steering 070°T at a speed of 12 knots sighted a lighthouse bearing 040°T. Half an hour later at 1120 the same lighthouse bore 356°T.

The course will already be laid off on the chart. The first bearing 040° taken at 1050 should be laid off from the lighthouse as a position line. Where it cuts the course line will be an estimated position.

From the lighthouse lay off the second bearing taken at 1120 so that it crosses the course line. The distance steamed by a 12 knot ship in half an hour would be six miles. From the first estimated position (1050) measure off six miles along the course line. If this measured distance places the ship on the second bearing at the second estimated position then the ship has made good a course of 070°T and has run six miles. The estimated course and distance has been accurate and no set or drift has been experienced.

If, however, the distance run of six miles, as *Fig 9.5* illustrates, takes the ship past the second bearing, then with the aid of the parallel rulers the first bearing should be transferred and drawn through the course line, at the position where the distance run ended. Where the transferred first bearing cuts the second bearing is the ship's position. Measure the distance back and reverse the course to prove the geometry of this method.

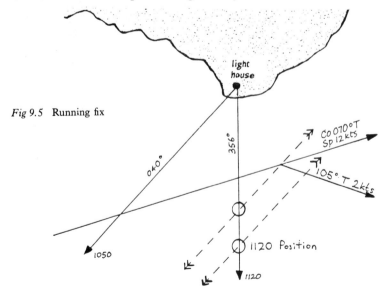

Fig 9.5 Running fix

If the navigator knows from his Admiralty Tidal Stream Atlas that a tidal stream of 105°, two knots was effective in this area, then after measuring off six miles for half an hour's steaming along the course line, a further allowance should now be made of one mile (half hour of tide) 105°T. The original bearing taken at 1050 of 040°T must now be transferred through this last position and where it cuts the second bearing is the ship's 1120 position.

Three cross bearings

So far we have dealt with fixes which have included the use of two bearings or a bearing and a sextant angle. It is much more satisfactory and accurate, however, to use three bearings and to find that they cross together or nearly together at one place. If the three position lines do not quite cross together the navigator will find that he has a small triangle on the chart, which is called a 'cocked hat'. The middle of the 'cocked hat' may be used as the position if it is small. If however the 'cocked hat' is large, the bearings should be re-taken and/or a check made to see if a mistake has been made in applying the compass error. (See *Chapter 8, Application of compass errors*).

If the navigator is in doubt about the identity of shore marks, three bearings will prove the accuracy of identity, or otherwise. By looking at *Fig 9.6*, it can be seen that bearings of a lighthouse and one of two beacons were taken about 90° apart, and a fix at position A was established. By taking a third bearing from a position marked Flag Staff, a new position B was fixed and it was found that the wrong beacon had been used in the first plot.

Two sextant angles – ranges

The use of a single vertical sextant angle and a bearing has been illustrated. By using two vertical sextant angles of different shore objects and following the procedure prescribed earlier, we will have two ranges or distances from two points. By looking at *Fig 9.7*, it can be seen that the distances off positions A and D will provide a fix, if the distances off are used as radii of circles. This technique may also be used with radar.

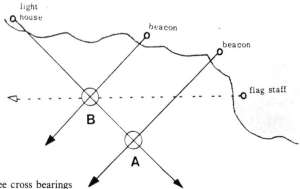

Fig 9.6 Three cross bearings

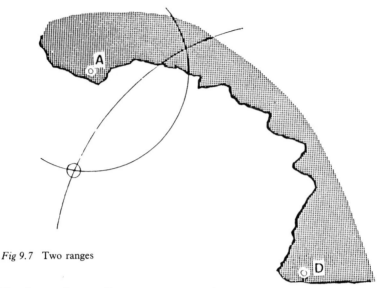

Fig 9.7 Two ranges

Bearing and sounding

When there is little difference in the heights of high and low water and where the soundings have a definite character and line, an approximate position may be found by using a single bearing and sounding. The sounding must have been corrected as described in *Chapter 5* and should lie along the position line. If, however, the soundings are uniform in depth, the chart in use has not been surveyed for many years, or the soundings are widely separated and not in lines, then this method of position fixing is not of value and may well be dangerous.

Angles on the bow

By doubling the angle on the bow, the navigator may fix his position by using the simple principles of geometry in the isosceles and right-angled triangle. Triangles all have three sides and three angles, the sum of which is $180°$. We know that in an isosceles triangle two sides are of the same length. It follows that if we have two sides of the same length then the opposing angles are of the same size in degrees. *Fig 9.8* illustrates that there is a ship at A steaming due east and there is a point of land two points ($22\frac{1}{2}°$) on the port bow. With the ship continuing to steam along the course line, we wait until the first angle at A has doubled and is now $45°$ on the port bow and note the distance steamed. If we look at the figure again we see that at position B, by taking $45°$ away from $180°$ we have angle $ABD = 135°$. Therefore the angle at D must equal $22\frac{1}{2}°$ and by doubling the angle of $22\frac{1}{2}°$ on the bow we now have an isosceles triangle. The sides opposite are equal and because we know how far we have steamed from A to B we now know the distance B to D is the same, which is the distance off.

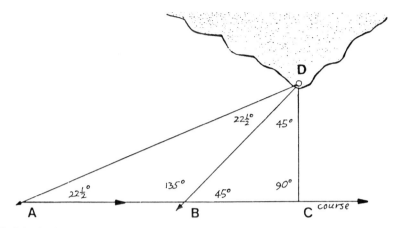

Fig 9.8 Angles on the bow

We now come to the 'four point bearing' an angle made at position B of 45° between the course line and point D. This is really a continuation of the isosceles triangle except that by doubling the angle of 45° we know in advance that our second angle will be the ship's position when we are abeam. The ship's position when abeam of a point of land is always of importance to the navigator; he now has an exact time and position of departure to his next point of interest and it is a good position to alter course. By taking the time and distance run from B to C we have formed a right-angled triangle BCD with the right angle at C. Angle at D must be 45° and because sides BC and CD are equal, we have a beam bearing and distance off.

Tidal effect on courses

By looking back to the section on 'Running fixes' it will be seen how a tidal stream affects the run of a ship when steaming. For the purposes of illustration and examination, the laying off of a tidal effect or current at the end of a run is acceptable. It is also reasonable to compare an expected position with a fix and name the difference as being due to speed and direction of current and wind. A navigator who knows the speed and direction of a tidal stream or current should allow for it before giving a course to steer so that the proper course can be made good.

If we look at *Fig 9.9* and suppose that a 15 knot ship wishing to make good a course of 090° is about to pass two islands, from between which there runs a current of 2½ knots in a direction 180°. From the departure point A, lay off the 090°T course which would be made good if the vessel steamed 15 miles in one hour and was unaffected by current to point B. Now lay off from A the departure point, the current's speed and direction expected over one hour, 180° – 2½ knots, to position C. Join C to B and the resultant course of 080°T is the course to steer in order to make good a course of 090°. *Note*: Always equate the ship's speed in knots to that of the current. The student may be told that a 15 knot ship is expected to be affected by a current running south at 2½ knots, over a distance of 90 miles when steaming east, and asked what course to steer to maintain an east course. The common factor is the 15 miles the ship covers in one hour and the 2½ miles the current runs in one hour.

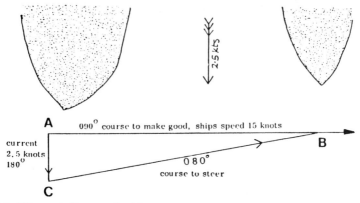

Fig 9.9 Effect and allowance for tide on course

Leading lines – clearing marks and clearing lines

We have already mentioned the use of a transit in giving a position line. Transits are also used to assist coastal navigation where two marks kept in transit lead the ship in the best channel. Such marks are called 'leading marks' and are shown on a chart by a line drawn through them called a 'leading line' (see *Fig 9.10*).

Clearing marks

When a hidden danger lies in the approach to a harbour or anchorage it will often be found that a 'clearing line' is on the chart. This line leads clear of the danger and a ship is in safe water if she keeps outside it.

Leading line and clearing lines

When having only one conspicuous object on which to run with dangers either side of a vessel's planned track into an anchorage or harbour, a line of bearing drawn on the chart and so marked in degrees from that object may be used to lead between and clear of them, with other lines also drawn and marked from the object to the limits of the dangers thus giving the limits of bearing within which the object must be kept; for example when steering courses to allow for the effect of a cross tidal stream (see *Fig 9.11*).

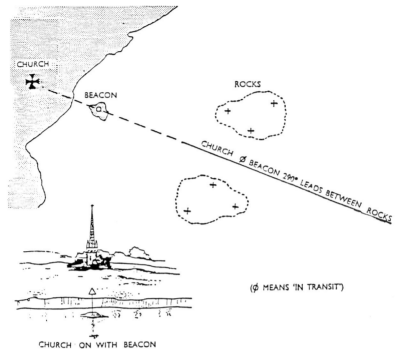

Fig 9.10 Leading marks and leading line

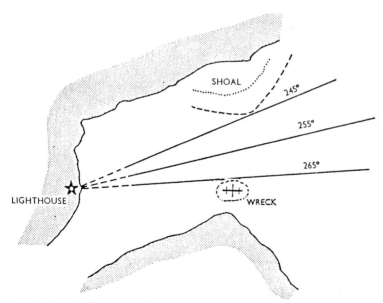

Fig 9.11 Leading line and clearing lines

10 Electronic navigational aids and equipment

Since its first introduction to the fishing industry, radio and electronic equipment has played an increasingly important part in the seeking, catching, and safe and economic landing of fish. The facilities this equipment offers have become an essential part of modern fishing operations and skippers and mates should be fully conversant with the proper operation, use and limitations of electronic aids.

Large distant water trawlers carrying radio officers usually have high power telegraphy sets capable of transmitting and receiving on high, medium and low frequencies. These sets may also be provided with a radio-telephone capability. Such equipment may be used for communication not only with coastal stations but with long range stations, *eg* Portishead.

Middle and near water vessels may not carry radio officers and for this type of vessel radio-telephone equipment is provided for the skipper or mate. There is a wide range of equipment that is simple to operate and efficient.

The installations on smaller near water vessels consist of medium power radio-telephone equipment with a built-in receiver and an extension speaker to the wheelhouse so that the officer of the watch is able to listen in on the calling frequency for distress or emergency calls. The speaker in the wheelhouse is also provided on fishing vessels which carry radio officers, to be used when the radio officer is not on duty.

Radio telegraphy and radio-telephone receivers are fitted with alarm signals which activate on the transmission of either a keying or two tone signal respectively. (See *Chapter 24*). The alarm signal gives the listener some warning that a distress message will follow.

Some receivers have an additional capability in that they are fitted with sensors and can be connected to a direction finding loop aerial. All these sets may be used to receive transmissions from National radio broadcasts concerning time signals, navigational/gale warning, weather forecasts, *etc.*

VHF radio

The international maritime VHF radio-telephone service is the most rapidly expanding facility available for relatively short-range interference-free efficient coastal communications. The UK, for example, has complete coverage around the coast out to about 30 miles to seaward. Its use by port operations, pilot services, tugs, Coastguards, lifeboats, *etc* creates an increasing volume of traffic which makes the VHF band the most versatile and widely used international service.

It must not be abused for unnecessary talk.

The silence period should be strictly observed on Channel 16, the calling safety channel. Skippers and mates should make themselves familiar with the procedure when calling coast stations by knowing the particular stations working channel in advance.

Consult Admiralty list of radio signals as follows;

Volume 1 – Coast radio stations, SAR procedures, medical advice by radio and AMVER. (Automated Mutual Assistance Vessel Rescue System).

Volume 6 – Stations working with port operations service, pilot service, and traffic surveillance

Radio direction finding

There are two ways of obtaining a RDF bearing. The more usual method is for the ship which is equipped with direction finding equipment to take a bearing of a lightship or lighthouse which transmits a distinctive and identifiable signal. The second method of obtaining a radio bearing is to request a coastal radio station to take a radio bearing of the ship (QTG) service.

In all cases the radio bearing will be a part of a great circle, which cannot be applied to a Mercator chart. However, the great circle bearing can soon be corrected for use on a Mercator chart. By taking the position of the shore station and the ship's DR position, the mean latitude and the difference of longitude can be found. In most nautical almanacs and nautical tables there will be half convergency tables. Enter these tables with the mean latitude and D longitude between station/beacon and ship, and take out the correction in degrees. Always apply the half convergency **towards the equator**.

	North lat	South lat
Bearing measured eastward	Add	Subtract
Bearing measured westward	Subtract	Add

Example. The navigator of a ship in DR position 58°30′N 2°30′E was given a bearing of 082° from Wick W/T DF station (58°25′N 3°10′W). Find the true bearing to lay on the chart.

Wick radio lat	58°25′N	Wick radio long	3°10′W
Ship's lat	58°30′N	Ship's long	2°30′E
Mean lat	58°27½′N	Diff long	5°40′

If using, for example, Burton's Table 38 for convergency angle and with mean lat $60°$ in the vertical column and diff long in the horizontal column, extract the correction of $2.6°$. The bearing was taken from Wick Radio $082°$; correct by bringing the easterly bearing towards the equator.

$$082°$$
$$+ \quad 2.6°$$
$$\overline{084.6°\text{from Wick Radio}}$$

This line of position may now be drawn 084.6T from Wick, and the ship is somewhere on the position line (*see Fig 10.1*). If the radio bearing had been taken from the ship it would have read

$$267.2° \quad \text{westerly towards equator}$$
$$\text{Correction} - 2.6°$$
$$\overline{264.6°}$$

Bearings of two or three coastal stations taken simultaneously will provide a good fix. If three can be taken which give a small 'cocked hat' when laid off on the chart, a reliable position is fixed.

There is, however, no reason why a radio bearing, which is a position line, cannot be crossed with another position line. If, for example, the observer was given a bearing from Wick Radio as in the preceding example and then took an observation of the sun, both position lines could then be laid off on the chart to give a fix.

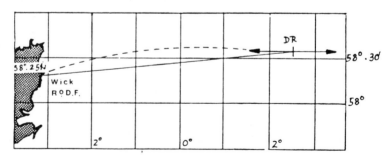

Fig 10.1 Position line obtained by DF bearing

Practical use of radio direction finder

There are two types of direction finding equipment, the fixed loop aerial and the rotating aerial. The former is the most reliable. *Disconnect any other aerial circuits at their sets if the aerials are near to the loop.*

Consult both chart and Admiralty list of radio signals in order to choose suitably placed stations that are within the range of reliability. At sunrise and sunset sky-wave errors are at their peak if the vessel is more than 70 miles from the station. At lesser distances reliable bearings may be taken by day or night, range depending upon output power.

Good bearings depend on the accuracy of the loop-reading, the accurate determination of the ship's heading at the time of the reading, and the range of the ship and station.

Adjust the receiver tuning for best reception on the transmitting station's frequency. When the station's call sign has been recognised, it will usually be followed by a long dash. Swing the pointer through the long dash until the sound fades altogether and on until the sound is just heard again. This silent arc is known as the null. Repeat the swing through the null, and the mean position in the null will be the bearing.

When the precise bearing has been ascertained check the precise course, because the bearing taken is relative to the ship's head. It must be stressed again that a $1°$ error when either taking the null bearing or noting the ship's head will result in an unacceptable error at a maximum range.

Radar

Nearly all fishing vessels are now fitted with radar and in some cases may be provided with two sets. Radar is an invaluable aid to navigation when used properly. It must be stressed, however, that radar has its limitations and officers who ignore plotting procedures and use radar for navigation with a casual or over-confident manner, may well find themselves in a disastrous situation.

We will begin by describing in simple terms how a radio detection and range (RADAR) set works. Radio pulses are transmitted at a very rapid rate of up to 4,000 pulses per second. When these pulses strike another ship, a buoy or land, some of the pulses will be reflected back towards the transmitter, from which they were sent. The aerial from which the pulses are sent and received is known as a scanner. The radio pulses are sent out in a narrow band in one direction only, but because the scanner revolves through $360°$ the horizontal plane is covered on each revolution.

The pulses when returned are measured on a time scale and cathode ray tube on to which they are shown as a visible echo. This is the radar screen and because the echo or trace is illuminated on the screen with an afterglow effect, repeated on each sweep of the scanner, a continuous picture may be obtained. It is important to mention here that an echo will only be shown or indicated on the screen if pulses are returned from the target to the scanner. If pulse return is weak or non existent, it will be obvious that, in poor visibility or at night, a situation could develop which might be dangerous, particularly in coastal and landfall navigation.

The range of the radar is almost the same as that of the human eye and it has a horizon which depends on the height of the scanner. Because the radar depends on a reflected signal it will clearly indicate hills or mountains which appear above the horizon. Low lying land, beaches or shoals may only produce a weak echo, and if they lie between the ship and the high land, the imprudent navigator may mistake the contour of the high land as being the coastline.

Similarly when visibility is poor in enclosed waters such as a buoyed channel, the radar will indicate the channel buoys clearly, but if the channel is narrow and the buoys are to be passed close to, the sweep of the scanner's pulse sent out from the top of the wheelhouse will pass over and above the buoy when closed. The target will not show on the screen, or it will be lost in the clutter or loom at the centre spot and a collision may occur. Most experienced seamen and pilots are aware of this situation.

A round lighthouse or a round Baltic type spar buoy designed to be conspicuously visible by eye will, because of its round shape, scatter the radar pulses and produce a weak echo. Rain and snow clutter distort pulses and fishermen should be aware of all the limitations under the conditions described above.

The quickly taken range and bearing of a point of land when navigating on the coast at night can be a positive danger if the observer is not absolutely sure that the point of land showing on the radar is identified and properly related to the course, distance, speed, and previous fix. Over-reliance on a radar picture and the neglect of established navigation procedures cannot be emphasised too strongly as being a dangerous practice.

A trawler bound from Hull to the south west coast of Iceland cleared the Pentland Firth and set course so that the Westman Isles were right ahead and with a distance of more than 400 miles to steam. The vessel was put on to automatic steering on the gyro-compass. The ship was fitted with a good modern radar set and it was expected that some 36 hours later the Westman Isles would be picked up. The weather was fine and clear with a fresh to strong southwesterly wind on the port side. All normal navigational procedures were then neglected. The Decca coverage from the Scottish chain, reliable for half of the run across, was not used to take an accurate fix or to determine leeway or set. The log was not set or used and the echo sounder was switched off. The sea was broken and choppy giving some degree of intermittent clutter. At about 0230 in the morning (36 hours later) a distinct echo appeared on the screen, right ahead, at a distance of 20 miles. Course was then altered to port so as to pass this target on the starboard side. It was presumed to be the Westman Isle. The trawler crossed the distinctive 100 fathom line which follows the contour line of the Icelandic coast, but the echo sounder was not in use. Soon afterwards the vessel ran aground on the low lying lava sands of Iceland and became a total loss. The echo had not been the Westman Island but a mountain peak, some eight miles inland from the coast, height 5,000 ft. The trawler had made leeway and had been set more than 25 miles on a 400 mile run. An error of 1½° on the gyro would have had a similar effect. Over-reliance and bad interpretation of the radar caused the loss of this ship. Fortunately no lives were lost. The reliance and confidence on this efficient radar was so great that distress signals were sent out saying that the vessel was aground on the southwest of Westman Island.

The lesson to be learnt from this incident is that *constant vigilance on all navigational procedures should be the watchkeeper's and skipper's priority.* They must not be neglected just because there is an efficient operational radar. The radar will only show echoes, which the navigator can identify, aided by previous fixes, soundings, distance run and a knowledge of leeway, set or compass error affecting the ship.

The next most common error is the double risk an observer may make by failing to plot the movement of another vessel in fog and also to assume that the plotted target has operational radar. If for example a target is picked up ahead, then *it must be plotted at regular intervals.* If its bearing does not change then risk of collision exists, and a large alteration of course should be made. The large alteration will be effective on two counts. If the other vessel has radar, the large alteration of course will very soon be apparent to the observer. If the vessel seen has no working radar, then one's own vessel will move away quickly from the close quarter situation.

Position fixing by radar

A radar position line may be obtained in one of two ways, by bearing and distance from a single point or by two or three ranges from two or three points. There is not much problem in making a fix by radar if the observer is very familiar with the local contours *eg* the home port. But in other places such as the Norwegian coast, with numerous rocky islets, high background mountains and inlets to fjords large and small, identity of points can be difficult. Parts of the Minch may be difficult to identify and under such circumstances with poor visibility the radar should be used to fix arcs of safe distance so as to keep off the land. Even in normal visibility, the most likely source of error will be to identify with certainty the exact position on the chart shown by the echo on the screen.

The accuracy of a radar bearing is always less than that of a visual bearing. The pulse signal transmitted by the scanner has an angular width of up to two degrees. As a result of this angle of 2°, measured at the scanner, the beam width increases in proportion to the distance off the target. Consequently a very small island or lightship, *etc* will appear as a blur on the screen. Similarly if the observer is unfortunate enough to have another target on approximately the same range and near to each other, then the echo of the one will be carried into the other. This can easily happen with a small archipelago of islands or cluster of dangerous rocks. An unchecked bearing marker may introduce further bearing inaccuracies.

Reflectors, ramarks, racons

It has been explained how radar pulses may reflect badly from curved lighthouses, flat low lying shores and buoys. Other targets such as wooden craft, ice, or fibre glass may reflect radar pulses badly or not at all. This can be overcome by the fitting of a radar reflector, except of course in the case of ice. Fishermen who are likely to have to fish or navigate in ice conditions are advised to read and study the *Mariners Handbook* published by the Hydrographer of the Navy. In this book there is a comprehensive description of ice conditions, navigation, and shiphandling in ice, as well as detection of ice by radar.

Radar reflectors are constructed so that they may be fitted to the top of the mast or wheelhouse of a wooden boat. They may be fitted to the top of a buoy or at each end of a wooden jetty, indeed on to anything which may be thought to reflect radio pulses badly. The reflector may consist of steel plates, intersecting with each other and set at different angles, so as to be able to respond readily to the radar pulse.

Ramarks

The ramark which transmits continuously from an important navigational mark, *ie* a lighthouse or lightship, will be shown on the radar picture as a continuous or dotted line. This line appears on the edge of the screen and runs towards the centre spot, and it gives the bearing but not the distance of the station transmitting the ramark signal.

Racon

The racon is a radar beacon which responds to the pulse of a radar transmission by transmitting its own pulse in a known pattern. The racon station's signal appears on the screen of the ship and the bearing is obtained. In addition to the bearing, the racon gives the radar observer the distance off and identity of the station. The signal begins close to the racon location and extends radially towards the outside of the display.

Radar types and use

There are two types of radar set and two kinds of display which may be obtained for each of the two types.

Unstabilised, ship's head up

On this display the heading marker always points to 000° on the degree scale ring surrounding the display. Bearings are always relative to the ship's head. The observer's position is always at the centre of the screen and the apparent track of another vessel will be the combined movement of the other vessel relative to one's own movement. A rapid alteration of course, *eg* swinging, will create shadow and temporary loss of definition.

Stabilised north up

By feeding the vessel's course from a gyro or magnetic compass into this display unit, the ship's head marker points to the compass course in degrees on the graduated scale surrounding the screen. The picture is always shown north up, with 000° on the scale being at the top of the screen. Bearings taken on the radar screen are compass bearings. The ease in taking bearings on this type of set is such that risk of collision with another vessel may be determined by the observer alone if he lays the bearing marker on to the target to see if there is any appreciable change. It is the most common type of display.

True motion north up

The observer's position on this display does not remain at the centre of the screen but moves according to course and speed across the screen. All stationary targets, ships at anchor, buoys and shoreline are in plan view and motionless, with other vessels as well as one's own moving across the screen on their different courses. When one's own vessel nears the edge of the screen it is automatically returned to commence another traverse. If the observer is concerned as to what hazards lie ahead, his own ship's position may be returned manually to the beginning of a traverse. The advantages of this type of set are high because all movement of other craft may be seen in a true format and collision situations appreciated more readily. On the true motion display, a feathered track or afterglow indicates the course of other moving echoes.

True motion head up

A radar set fitted with stabilisation capability may be adjusted so that the ship's course is head up instead of north up. In pilotage waters there is an advantage in that other vessels ahead and astern can be seen in relation to the narrow channel and the buoyage therein.

Control settings

Whichever type of display unit is available, the user must be able to recognise loss of performance and maladjustment of the controls. The correct settings can be recognised as follows:

Brilliance. The rotating trace should be only just visible, this may require re-adjustment when changing range. The gain should be turned off when adjusting brilliance control.

Focus. The range rings should be as thin as possible.

Heading marker. This should point to $000°$ on the ship's head up display and to compass course on stabilised displays.

Sensitivity (gain). Correct adjustment is easily recognised on the longer ranges by the speckled background being only just visible.

Anti-clutter. Ideally this control should be set so that the sea clutter is only just visible. In practice it will probably be found that this control needs continuous adjustment so that too dark and too bright areas may be examined.

Performance monitor. The performance monitor is the only real sure means of checking that both transmitter and receiver are working properly. The watchkeeper should be familiar with the length of plume or sun effect which is displayed when the monitor is switched on.

Rain switch (differential). This control sometimes helps to pick out targets in an area of rain, but increased gain may be required when it is in use.

Use of radar

The International Regulations for Preventing Collisions at Sea (1972) take the use of radar into account.

Collisions have been caused far too frequently by failure to make proper use of radar; by altering course on insufficient information and by maintaining too high a speed particularly when a close quarters situation is developing or is likely to develop. *It cannot be emphasised too strongly that navigation in restricted visibility is difficult and great care is needed even though all the information which can be obtained from radar observation is available. Where continuous radar watchkeeping and plotting cannot be maintained even greater caution must be exercised.*

Clear weather practice

Whether or not radar training courses have been taken it is important that skippers and others using radar should gain and maintain experience in radar observation and appreciation by practice at sea in clear weather. In these conditions radar observations can be checked visually and misinterpretation of the radar display or false appreciation of the situation should not be potentially dangerous. Only by making and keeping themselves familiar with the process of systematic radar observation and with the relationship between the radar information and the actual situation, will officers be able to deal rapidly and competently with the problems which will confront them in restricted visibility.

Interpretation

— It is essential for the observer to be aware of the current quality of performance of the radar set (which can be most easily ascertained by a performance monitor) and to take account of the possibility that small vessels, small icebergs and similar floating objects may escape detection.
— Echoes may be obscured by sea or rain clutter. Adjustment of controls to suit the circumstances will help, but will not completely remove this possibility.
— Masts and other obstructions may cause shadow sectors on the display. These sectors should be assessed and recorded on a card on the bridge.

Plotting

To estimate the degree of risk of collision with another vessel it is necessary to forecast her nearest approach distance. Choice of appropriate avoiding action is facilitated by knowledge of the other vessel's course and speed, and one of the simplest methods of estimating these factors is by plotting. This involves knowledge of one's own ship's course and the distance run during the plotting interval.

Appreciation

— A single observation of the range and bearing of an echo can give no indication of the course and speed of a vessel in relation to one's own. To estimate this a succession of observations at known time intervals must be made.
— Estimation of the other ship's course and speed is only valid up to the time of the last observation and the situation must be kept constantly under review, for the other vessel, which may or may not be on radar watch, may alter her course or speed. Such alteration in course or speed will take time to become apparent to a radar observer.
— It should not be assumed that because the relative bearing is changing there is no risk of collision. Alteration of course by one's own ship will alter the relative bearing. A changing compass bearing is more to be relied upon. However, this has to be judged in relation to range, and even with a changing compass bearing a close quarters situation with risk of collision may develop.

Operation

— If weather conditions by day or night are such that visibility may deteriorate, the radar should be running, or on 'standby'. (This latter permits operation in less than one minute, whilst it normally takes up to five minutes to operate from switching on). At night in areas where fogbanks or small craft or unlighted obstructions such as icebergs are likely to be encountered, the radar set should be left permanently running. This is particularly important when there is any danger of occasional fogbanks, so that other vessels can be detected before entering the fogbank.

— The life of components, and hence the reliability of the radar set, will be far less affected by continuous running than by frequent switching on and off, so that in periods of uncertain visibility it is better to leave the radar either in full operation or on standby.

Radar watching

In restricted visibility it is always best to have the radar set running and the display observed, the frequency of observation depending upon the prevailing circumstances, such as the speed of one's own ship and the type of craft or other floating object likely to be encountered.

Radar training

It is essential for a radar observer to have sufficient knowledge and ability to recognise when the radar set he is using is unsatisfactory, giving poor performance or inaccurate information. This knowledge and ability can only be obtained by a full and proper training; experience alone or inadequate training can be dangerous and lead to collision or stranding through failure to detect the presence of other vessels or through misinterpretation of the radar picture.

Radar in narrow channels and port approaches

In poor visibility, whether a pilot is carried or not, as much information on ships at anchor or moving should be obtained from the port information service. If the fairways and approaches are clear it may be considered safe to proceed, but it is not safe to leave a dock or proceed to a dock with a strong ebb or flood tide running in the same direction as the ship. It will be obvious that a vessel proceeding against the tide is under better control in poor visibility than the vessel with the tide astern, whether the channel is clear or not. To proceed up river in fog on a strong flood tide, or to sail down river from a dock on an ebb tide is to invite trouble. In an emergency, complete control of the single screw ship will be lost by going astern when running with the tide.

The echometer

Originally produced as a navigational aid by showing the depth of water beneath the vessel between the sonar transmission points and the bottom, the echometer soon proved itself capable of indicating the presence of fish. This latter facility was pursued in the design and development of later models and unfortunately the echometer is now regarded by many fishermen as being primarily an aid to fish catching and they neglect to use the machine as an aid to navigation. Most large fishing vessels are fitted with two sounders, the electronic display and the recorder, which traces the depths in a pattern on a paper roll. The graph shown on the paper roll depicts the contour of the sea bed and consequently indicates wrecks or other fasteners on which fouling of gear may occur.

The echo sounder works by reason of an underwater sonic transmission emitted from the bottom of the ship, so that when reflected from the ground it is received at the bottom of the ship and the time interval is measured. The speed of sound through sea water is known for varying temperatures and salinity. Normally it is 1500 metres per second, thus the time interval is converted to distance in metres. Continuous signals are shown as continuous depths and may be displayed so as to be visually observed, or they may be shown on a sounding trace, to be seen as well as recorded on paper.

In shallow water with only a few feet under the bottom, an echo sounder may show greater depths than actual by reason of the signal reflecting several times between the ship's bottom before being received and transferred into a distance. For dock and harbour soundings special echometers capable of sounding to less then a foot are available.

A more recent development in the echo sounding equipment is that of the capability of selecting a particular band of soundings, ie 150 fathoms to 200 fathoms. This band selection is of particular use to fishermen in that the band chosen is expanded to be shown on a longer scale. Consequently more detail of the bottom can be seen and of course fish associated with bottom trawling can be more easily seen. Similarly a band can be selected when fishing in the pelagic mode, regardless of depth of bottom, so that pelagic fish shoals may be seen. It is important to remember to return the echo sounders mode of sounding to normal when fishing is completed and it is intended to begin steaming.

Decca navigator

The Decca receiver, now widely used on nearly all ships navigating northwest Europe and areas mentioned in Chapter 6, has a capability of receiving radio signals from a group of stations and converting them into readings on dials. The readings, identified by colour as well as numerically, may be read off at any instant and plotted on a Decca lattice chart. The accuracies of position fixing are good and more than enough for coastal navigation.

The UK and continental coasts are covered by groups of stations for different areas known as chains. Each chain has a master and three slave stations, each Master/Slave gives a hyperbolic pattern and position line which can be referred to the chart quickly and accurately for position fixing. Lane identification is provided so that the receiver can be set up and charts are marked according to the chain for which they are latticed. The range of reliability is for about 240 nautical miles from the master station.

There is a Decca Marine Automatic Plotter which, when linked to the Decca Navigator, provides a visual record of the ship's progress on paper. This is a most useful facility when fishing on the edge of a bank and when on fish.

When navigating in poor visibility, homing along lanes towards a particular point such as a light vessel or fairway buoy can be dangerous in that other vessels may have adopted the same procedure on similar or reciprocal courses.

Loran

Long range aid to navigation systems (Loran) are now being more widely used on distant water fishing vessels. As these systems have been developed, a greater accuracy in position fixing has been achieved. Pulsed hyperbolic signals are transmitted from a master station and slave stations and they are time measured by the receiver on board a ship.

The relative time difference (TD) between the signals sent out from the Master/Slave No 1 station is represented by a hyperbola and can be put on a chart as a position line. Similarly the measured time difference between the signals sent out simultaneously from the Master/Slave No 2 station give a second position line. Where these two position lines intersect is the position of the receiver. Reference to a Loran chart which is specially marked with these hyperbolic lines of position, usually 100 microseconds apart, establishes the geographical position.

Both systems Loran A and its successor Loran C utilise pulsed hyperbolic signals. Loran A stations send out a signal pulse. Each Loran C station sends out a group of eight signal pulses. This is done to improve the signal-to-noise ratio and thereby increase range and accuracy.

Loran A stations send out a signal pulse and receivers use only the pulse envelope to measure the time difference; as a result accuracy and fine resolution inherent in the Loran C system cannot be achieved. Accuracy of position is from 200–2,000 yards, range 700 miles max.

Loran C chains have several slave stations but the set is limited to provide a readout of two position lines so that the two that provide the best intercept are selected for display. Loran C receivers first measure the pulse envelope, this is called pulse matching, and obtain an approximate time difference reading. They then use an RF carrier to achieve fine resolution, this is called cycle matching, and in ground wave areas at a distance of at least 1,000 miles accuracy of fixes can be within 50–500 yards.

Full service Loran C sets automatically track and display two time difference readings, either simultaneously, or alternatively every few seconds. Watchkeeping fishing vessel officers, before using Loran C, should determine from the Loran C chart which chain is appropriate for their approximate position and set the pulse rate code for this chain on the chain select switches. Switch on and allow ten minutes for the set to stabilise. In good ground wave cover the set should lock on to the appropriate stations.

Further adjustments for signal interference, extended range and station change can be made easily, and watchkeepers should study the operator's manual.

Skippers and mates having switched on the Loran C set in plenty of time should check the position obtained against a known position taken from Decca Navigator or shore bearings before leaving the land or passing beyond the Decca Navigator range of reliability.

They will then know that the set is synchronised and they will appreciate that it is important for fishing vessels to obtain a high degree of accuracy in position fixing, which enables them not only to fish efficiently but to keep clear of closed fishing areas enforced by the government of other nations.

Automatic plotters are available for use with Loran C receivers. If, at night or on extended range, the ground wave signals are not being received properly, the Loran C set will continue to track on sky wave and give out readings. Depending on the make of the set, warning lights will be shown on the display unit indicating that sky wave is in use and that positional accuracy has deteriorated.

Watchkeepers should not work to fine limits under these circumstances because positions may be inaccurate by a greater amount than expected.

Omega worldwide navigation system

Omega is a system which measures the phase difference of very low frequency radio waves. From eight appropriately sited radio stations around the world, transmissions are made which give global coverage. The stations designated from A to H can provide a ship with digital lane information for use on an Omega chart from any two stations anywhere in the world, by night and by day with an accuracy of up to two miles.

Because the signals from each station cannot be discriminated if they are transmitted simultaneously, each of the eight stations sends out the signal having the same phase but at separate times in the transmitting sequence at 10 second intervals.

Before using Omega the starting position ± 4 miles must be entered into the receiver which then tracks automatically on the appropriate predetermined stations. Some Omega receivers have a provision for a supplementary third line of position which may be used as a cross check on the accuracy of the first two lane readings.

There is an error which must be applied to Omega readings which is known as a propagational correction. This error is caused by the very low frequency reflecting layer (ionosphere) changes in height, as by day and night, so that a small difference in the readout position line and the actual position line occurs. This propagation error may be entered into the set so that a true readout of position may be obtained from the receiver and the error to apply is obtained from propagation correction tables.

The Omega receiver operation is designed to be simply and quickly mastered, even by the inexperienced, when used in conjunction with Omega charts.

Satellite navigation

In orbit around the earth are five navigational satellites. From each satellite, data is transmitted defining the orbit with accurate time markings. This information, both orbital and time, is updated by the ground stations. The ground stations track the orbiting satellites constantly and any changes in the satellite's orbital parameters are put into effect by re-transmissions from the satellite.

The satellite receiver on board ships locks on to a satellite when it appears on the horizon; accepts orbital information, time, and doppler shift; and computes this data to produce an exact position in degrees, minutes, and decimals of a minute in terms of latitude and longitude. No corrections are necessary.

Between satellite passes the computer will supply a continuous display of DR positions if course and speed are input. On the next satellite pass the set will update the DR position with a new positional fix.

Corrections are necessary for the height of the ship's aerial above the Earth's theoretical surface. The indicated position will be in error if an incorrect course or more particularly, an incorrect speed are entered.

Accuracy of the indicated positions should be within ¼ mile when a satellite is being observed.

Electronic charts

A development in marine navigational equipment which is receiving increasing attention is the so-called 'electronic chart'. The term 'electronic chart' usually conveys an integrated navigational system which combines on a VDU a simplified chart with position information, and preferably also with a radar image. Such systems should perhaps more correctly be called electronic navigation systems. Several makes of equipment, most of which are in only early stages of development, are already available. Currently there are practical difficulties such as the provision and updating of reliable chart data. In theory, users will have freedom to select such chart data for display on whatever scale they consider most relevant to them, but errors in making these choices could have serious consequences. Many problems remain to be overcome before electronic navigation systems equal paper charts in adequacy and reliability. Electronic charts should not be considered as the equivalent of conventional charts and the carriage of paper charts remains mandatory under International Conventions (1974 SOLAS).

11 Ocean navigation

Astronomical navigation

When able to establish their position out of sight of land by electronic navaids, skippers and mates are naturally reluctant to do so by celestial navigation, which they consider time consuming in their already demanding routine. Not many of them have occasion to make long ocean passages and they therefore get out of practice in what is actually quite a simple mathematical process. Moreover the areas in which they operate are often subject to overcast weather. Nevertheless navaids can fail due to power failures and other causes and it is therefore necessary for those in charge of the larger fishing vessels and those making ocean passages to have a practical understanding of astro navigation, both to be able to establish their ship's position and to determine compass error by observation of the bearing of the sun or other heavenly body.

Whilst a knowledge of the theory helps, well established methods of calculation to a routine pro-forma will enable a skipper to achieve his object quite simply: all he needs is a sextant, a watch accurate to the nearest second or so and a nautical almanac and tables. The process is further simplified by the use of rapid method sight reduction tables and also the availability of electronic calculators.

The theory of Nautical Astronomy may be studied in various textbooks, in particular the 'Navigation Primer for Fishermen' referred to in the list of useful publications at the back of this volume.

The celestial sphere, or concave, is the heavens or space which surrounds the earth. For practical purposes, we assume that it is a hollow sphere, with all the heavenly bodies being on its surface and the earth at its centre.

Angular measurements on the celestial sphere correspond to those on the earth, the celestial equator being on the same plane as the terrestial equator and north and south latitude corresponding to declination on the celestial sphere. Longitude on the celestial sphere is termed Hour Angle and is the angle at the pole measured from the meridian of Greenwich westward from $0°$ to $360°$.

By using a Dead Reckoning (DR) or Chosen Position the taking of a sextant altitude of the sun will, with the use of tables, give the observer a position line.

The simplest position line to establish in this way is one's latitude by observation of the sun's meridian passage or by observation of the Pole Star. Fishermen operating in northern waters, with long periods of twilight, will find Polaris to be of the greatest value in finding latitude quickly and accurately.

Because Polaris is less than $1°$ from the celestial pole at any time, a small correction taken from the Pole Star Tables and applied to altitude taken by sextant will give the latitude.

Checking compass

At sea the easiest way of determining compass error is by taking a bearing of the sun at sunrise or sunset, when its bearing is known as its Amplitude. To allow for refraction, Amplitude should be taken when the sun's lower limb is about half the sun's diameter above the horizon.

Knowing one's approximate latitude and the sun's declination from a nautical almanac the true bearing of the sun can be found from any comprehensive almanac or nautical table and the error of one's compass thus determined.

Plane sailing – use of traverse tables

In the chapter on charts we have already seen how an approximation to Great Circle sailing on long ocean passages can be achieved by transferring convenient sections from the straight course on a Gnomonic chart (*ie* Great Circle) to a Rhumb line course on a Mercator chart, which normally serves for practical purposes.

Further help is readily available to navigators in keeping their dead reckoning (DR) at sea in Rhumb Line sailing (Plane sailing) by the use of Traverse Tables, to be found in Burton's or Norie's Tables.

These tables give solutions to right-angled plane triangles for every degree from $0°$ to $90°$ and for units of distance up to 600 miles. The variable factors in all such triangles are difference of latitude, difference of longitude, departure (distance east or west), course and distance. Departure may be converted into difference of longitude (D.Long) and vice versa by entering the tables with the Mean Latitude as the angle, the Departure in the D.Lat column and the D.Long in the distance column.

The Traverse tables enable one to determine, among other things, an onward position given course and distance from a known position; and also a course and distance between two given positions for distances up to 600 miles – beyond this the curvature of the earth makes this method inaccurate and a formula known as Mercator sailing must be used.

Mercator sailing

This is used principally to find the course and distance between two positions on a Mercator chart by taking account of the expansion of the latitude scale.

This method is important in high latitudes or when the differences in latitude and longitude are so large that Plane sailing methods are insufficiently accurate.

Mercator sailing is based on the system of Meridional Parts, simple formulae of trigonometry and the use of logarithms.

We have seen that on a Mercator chart the distance between parallels of latitude increases proportionately between the equator and the poles. The meridians instead of converging towards the poles are represented as straight parallel vertical lines a constant distance apart.

Meridional parts (MP) are a tabulation to determine the length of a meridian from the equator to a specified latitude expressed in minutes of the longitudinal scale. These may be found in Burton's, Norie's and other books of tables.

The difference of meridional parts (DMP) is the number of minutes of longitude measured along a meridian between two parallels of latitude. From the tables we find the MP of any latitude and can therefore calculate the difference of meridional parts between two latitudes.

The solutions to problems of mercator sailing can then be found by the use of two formulae; where D.Lat is difference of latitude and D.Long difference of longitude (in minutes):

— $\dfrac{\text{D.Long}}{\text{DMP}}$ equals tangent of course.

or Log. D.Long minus Log. DMP equals Log. tan. co.

— $\dfrac{\text{Dist}}{\text{D.Lat}}$ equals secant of course.

or

— Log. dist equals log. D.Lat plus log. sec. co.

The theory and examples of Plane and Mercator sailing are fully set out in the 'Navigation Primer for Fishermen' already referred to.

Part III – Watchkeeping, shiphandling

12 Officer of the watch

The officer of the watch is responsible for the safety of the ship and the lives of all those on board, whether the ship is under way, on passage, at anchor, or moored in a harbour with watchkeeping maintained.

Skippers of fishing vessels must understand that they are at all times responsible for the safe handling, navigation and management of the vessels in their care and that they may be called to account for any mismanagement or misdemeanours of themselves or crew.

It is in the skipper's own interests to give every facility to his officers to read *Notices to Mariners*, National Notices (such as UK 'M' Notices) and instructions from owners which may concern them. Deck officers should always have access to charts, nautical books/documents and equipment. Skippers and their officers must be fully conversant with the IMO booklet *Recommendation on Basic Principles and Operational Guidance Relating to Navigational Watchkeeping*.

The skipper of the fishing vessel is therefore bound to ensure that the watchkeeping arrangements are adequate for maintaining a safe navigational watch. Under the skipper's general direction, the officers of the watch are responsible for navigating the ship safely during their periods of duty when they will be particularly concerned to avoid collision and stranding.

Before taking into consideration the composition of the watch, which may vary with circumstances, it will be necessary to organise a watch system in such a manner that the efficiency of the watchkeeping members is not impaired by fatigue. Therefore, the duties should be so organised that the first and subsequent relieving watches are rested and fit prior to going on duty. It is well known that fishing vessels, because of heavy fishing, may have had an officer on deck for a long time if he was the officer responsible for the care and stowage of the fish. It would be wrong therefore for a man in these circumstances to have to take over a long spell of watchkeeping when tired, simply because it happened to be his watch. The skipper, under such circumstances, would have to make alternative arrangements. The care and attention given to the catching and stowing of the fish would count for nothing if the ship were to be put into danger by a tired watchkeeper.

When deciding the composition of the watch the following points should be taken into account:

110

- The officer of the watch should be a certificated and/or a competent person capable of complying with the *Regulations for the Prevention of Collisions at Sea.*
- At no time should the bridge be left unattended.
- Weather conditions, visibility, whether darkness or daylight, traffic congestion.
- The proximity of navigational hazards which may make it necessary for the officer in charge to carry out additional navigational duties.
- The use and operational condition of navigational equipment such as Radar, Decca, Loran, echo sounder, gyro or any other equipment likely to affect the safe navigation of the ship.
- Whether the ship is fitted with operational automatic steering.
- Any additional demands on the navigational watch which may arise from special circumstances.

Taking over the watch

The officer of the watch should not hand over the care of the ship to the relieving officer if he has reason to believe that the latter appears to be under any disability which would preclude him from carrying out his duties.

The officer taking over the watch should only do so when his vision has fully adjusted to conditions of darkness. Standing orders, special night orders or instructions left by the skipper in the night order book or deck log should be read and understood by the relieving officer, who should then satisfy himself on the following points:

- Position, course, speed and draught of the vessel. Gyro and compass errors.
- Direction and speed of tides, currents, weather, visibility and effect on the ship.
- Landmarks and/or seamarks in sight or expected to come into sight.
- Vessels in sight, their movements and bearings.
- The lights and signals being shown.
- Navigational aids in use and operational.
- State of weather and that expected; hazards likely to be met during the watch.
- Charts in use and those to be used readily available.

If at the time when watches are being changed, an alteration of course or other action is being taken to avoid a hazard, then the changeover must be deferred until the manoeuvre has been completed.

Keeping the watch

The watchkeeping officer should, whenever possible, fix the ship's position at regular intervals by whatever means possible. When coasting, fixes should be taken and laid off hourly, with the fixes being made by more than one method. Log readings should be taken at the time of the fix. Any set or drift away from the course to be made good should be allowed for by adjustment of the course steered.

The compass error must be checked regularly when conditions permit and taken into account in an alteration of course. Bearings of approaching vessels should be taken frequently in order to establish the existence of collision risk, with early and positive action taken as required by the Collision Regulations.

By regular position fixing, both by compass and/or electronic aids, with regular log readings and course adjustment, the officer will conform to good navigational practice, which will enable him to pick out and positively identify landmarks/seamarks as they are approached and come into sight.

Similarly the good watchkeeping officer when on duty and expecting to sight a landmark, lightship or buoy, will inform the lookout of the characteristics and approximate bearing of the mark expected to be seen.

In clear weather, when circumstances allow, the radar should be switched on and used for practice and expertise which can be evaluated against visual fixing, bearings and collision risks.

Restricted visibility

If the officer of the watch knows that visibility is deteriorating and suspects fog conditions, it is his duty to comply with the Collision Regulations and take the following action whenever possible before the fog is encountered:

— Call the skipper
— Post look-outs and helmsman (if on automatic steering)
— Proceed at a safe speed, sound fog signals, exhibit navigation lights
— Warn the engine room, be prepared for manoeuvring, man the bridge controls if fitted
— Operate and use the radar before entering the fog, to detect other vessels which may be unsighted.

When proceeding parallel to a coast, a course should be chosen which leads the vessel away from, rather than towards, the land. Turning points at lightships or headlands should be approached with caution. Fairway buoys, port approaches and narrow navigable waters such as the Dover Straits and Pentland Firth are convergency areas with high traffic density where the utmost skill and caution must be used in poor visibility in order to avoid collision.

Lookouts

In fog or reduced visibility a second and additional lookout should be posted so that the officer of the watch may give his full attention to the navigation of the ship. The word 'lookout' is now defined in the Collision Regulations as being a term which means a lookout being kept by sight, by hearing and any other available means. The latter term can be applied to a radar lookout.

When the skipper has been called for any reason at all he may come to the bridge and assess the situation. When fully aware of the conditions, he may take over from the officer of the watch. There should never be any doubt between the officer of the watch and the skipper when such circumstances arise. When the skipper feels that he has evaluated the situation, when all data and information has been given to him and he is ready to take over then he must so inform the officer of the watch. Experience shows that accidents have taken place when the skipper came to the bridge because the officer of the watch assumed that the skipper was in charge immediately, even before he had time to become orientated with the situation and when no clear hand-over took place.

Calling the skipper

The officer of the watch should call the skipper immediately when the following circumstances arise:

— If visibility deteriorates or is suspected to deteriorate.
— In traffic congestion or if other vessels are causing concern.
— When there is difficulty in maintaining course.
— Failure to sight land or navigation marks as expected. An unexpected change in soundings.
— If land or navigation mark is seen unexpectedly.
— Any breakdown of engines, steering or essential navigational equipment.
— In heavy weather, where doubt exists on damage being caused by taking seas, pounding or engine racing.
— If in any situation which gives rise to any doubt whatsoever.

In consideration of the above circumstances the officer of the watch should not hesitate to take immediate action in the interests of the safety and care of the ship, *eg* if pounding heavily or if heavy seas are coming on board. Similarly if an unexpected mark or signal is seen ahead then the safest action to take at once is to stop or reverse course immediately.

Deck log books

All fishing vessels are supplied with a deck log book in which a continuous account is kept of the ship's progress and position. The log covers the whole of the trip from leaving the berth to returning to berth in the home port. In addition to the navigational information entered in the log book, a record of the weather, fire and other safety drills and rounds is entered as well as any other incidents concerning crew or other vessels.

All incidents occurring during a fishing trip, including incidents on the fishing grounds, such as casualties to personnel or vessels, however slight, losses of gear, breakdowns, or damage to property must be fully written up immediately after the incident giving time, position, compass heading, engine and wheel movements and any other relevant data.

Deck log books must be properly written up during and at the end of each watch by the OOW. These log books are very important to owners and skippers, they are called in evidence at all official and other inquiries.

Notes on keeping the log

— The wind, sea and visibility columns must be filled in at the end of each watch in accordance with accepted scales. (*ie* Beaufort).
— In bad weather, take and enter the barometer reading hourly. Note the change. At other times enter the barometer reading at the end of each watch.
— Enter the course to be made good, true course steered, gyro and magnetic with error of deviation and variation.
— Enter the actual log reading every hour and at course and speed alterations.
— The OOW must initial the log book when going off watch. By so doing this will avoid considerable mental effort later when trying to ascertain who was on duty at a particular time.
— As before stated the log constitutes a continuous record and must be fully kept up when on passage. When not on passage, *ie* when fishing, a continuous record of weather conditions must be entered. The 'Remarks' column should contain details of other activities such as 'fishing', 'dodging', 'alongside', 'at anchor', *etc*.
— Where a fire detection system is fitted, the OOW should check the annunciator panel at the same time as fire and integrity rounds are made. Enter in log book.
— Erasures should not be made in the log book. If it is necessary to make an alteration, rule out and initial.
— When going on duty, the OOW should see that the previous OOW has initialled the log book and should check to see if the skipper has written in any night orders at the foot of the page in the space provided.

The officer of the watch, when either on passage or at anchor, must always remember that the engines and the ship's whistle or siren are there for his use as well as the steering gear. In the majority of cases risk of collision and stranding may be avoided by a timely alteration of course but the OOW should not hesitate to use engines and whistle signals in case of need. He should also keep in mind the manoeuvring capabilities of his ship, *ie* turning and stopping distances.

At anchor

When a fishing vessel is at anchor it is normally considered necessary to maintain bridge and engine room watches as if the vessel were at sea. The following procedures should be observed:

— Determine and plot the ship's position by visual bearings and plot on the largest scale chart. When possible, note shore or navigational marks in transit, for speed and convenience of checks.
— Note the time of HW/LW, strength and direction of flood and ebb. Note soundings on echometer.
— Take radar bearings and distance of prominent and identifiable shore mark when visibility is good. Note the readings as they may become useful if fog develops and the vessel remains at anchor.
— Maintain a proper lookout
— Maintain fire and integrity rounds.
— Call the skipper if the vessel is thought to be dragging the anchor. Use engine and steering as necessary.
— Call the skipper in the event of fog developing and comply with fog signals applicable for vessels at anchor in the Collision Regulations.
— Exhibit appropriate lights and shapes for a vessel at anchor in the Collision Regulations.

OOW with pilot embarked

Despite the duties and obligations of a pilot to both the ship and her owners, his presence on board does not relieve or exonerate either the skipper or the OOW from his duties and obligations for the safety of the ship. They should co-operate closely with the pilot and keep an accurate check on the vessel's position and movements.

If there is any doubt as to the pilot's intentions or actions, clarification should be sought from the pilot, and if doubt still remains, the skipper should be called immediately and action taken at once as the OOW or the skipper considers necessary.

OOW and personnel

The OOW should give the members of his watch all appropriate instructions and information which will ensure the keeping of a safe watch, a proper lookout and regular fire and integrity rounds of the vessel.

When men are working either on deck or aloft, the OOW must remember that he is responsible for their safety. If weather conditions worsen so that the movement of the ship is likely to create a risk to men aloft, or if shipped water becomes a hazard to men working on deck, then the OOW should not hesitate to bring the men down from aloft or in from the open deck. Lifelines should always be rigged when considered necessary.

13 Shiphandling

Shiphandling in port and at sea

Within the meaning of the word shiphandling, there are two basic practices, *ie* handling a ship in narrow confined waters, such as a dock, tidal or non tidal river or estuary, channel, *etc* and, secondly, the handling of a ship at sea in all weathers. It is therefore proposed to divide shiphandling into two parts as described.

Shiphandling in harbour

The art and skill of shiphandling varies to a large degree on the type of vessel one is called upon to handle. Experience in shiphandling is a valuable asset.

There are different degrees of skill required for different ships. The twin screw vessel with a bow thrust is probably the easiest type of vessel to handle. This type of vessel is usually a passenger ferry, oil rig or perhaps a naval vessel and it will usually be a high-powered vessel. At the other end of the scale there is the large single screw vessel with a large cubic capacity and low power. When heavily laden, these vessels with a low power to weight ratio are the most difficult to handle, both at sea and, particularly, within the confines of a port. Because trawlers and other fishing vessels are mainly single screw vessels without bow thrusts, we will describe only the handling of such vessels, on the assumption that if one is capable of handling a single screw ship, then the technique of handling a twin screw vessel or one fitted with a bow thruster will prove simpler.

Much has been written and illustrated on shiphandling which is largely geometrical and theoretical. Let it be said that the basic principle involved in successful shiphandling is that a vessel should have as little headway as possible when swinging or on the approach run to a lock, berth or jetty. Too much headway when manoeuvring is the cause of nearly all shiphandling accidents. In most cases the skipper will realise that there is too much headway on his ship when the lock or quay is neared. By going astern in a single screw ship in order to take off the headway, all control will probably be lost; the ship may cant violently and a collision with the quay or lock will be inevitable. Judgment of a vessel's headway plays an important part in assessing movement. The normal method of judging a ship's speed or headway prior to manoeuvring is to look at the shore marks abeam to see how quickly the ship is passing them. However, if the ship is several hundred feet away from the shore it would seem that the vessel is hardly moving. But if the ship is within a hundred feet of the bank or shore it will appear to be passing the marks very quickly. Judgment of a ship's speed by eye can be related to the distance of ship from shore. It is advisable, if possible, to run all the way off or go astern, before an approach run is made directly to open lock or jetty.

If the reader refers to *Chapter 5* it will be seen that the tide runs strongest between three hours and two hours before high water, reducing in velocity down to zero knots at high water. It is prudent therefore to approach and manoeuvre off a tidal dock as near to high water as possible. The way cannot be taken off a ship which is proceeding up river on a strong flood tide; consequently direct entry into a lock or the swinging of a vessel under such circumstances (depending on swinging room) will be difficult.

The basic lessons to be learnt are:

— Do not begin any manoeuvre with too much way on the ship. If in doubt stop the ship and begin from a stopped position.
— Avoid an approach to a dock, harbour or swinging area on a strong flood tide and where space or distance is limited.
— Learn and practice the judgment of speed over the ground on approach runs on slack or non tidal water. When approaching a jetty, try to stem the tide so that headway may be easily adjusted.

Remember that the quickest manoeuvre is that which is carried out slowly.

Transverse thrust, single screw

In the previous description of approach to a lock, it was stated that by having too much headway on approach to a lock or jetty, making it necessary for the engines to be put astern, control of the ship might well be lost by the vessel canting to one side. This sheering effect is peculiar to single screw vessels and is known as transverse thrust. Conventionally, ships have been fitted with engines so that the shaft and propeller, when turning ahead, rotate in a clockwise direction, when looked at from astern.

If we look at a right-handed propeller from aft which is going ahead (*Fig 13.1*), it will be seen that a greater pressure of water on the ship's hull is engendered by the blade in the top position and a weaker effect follows the blade at the lower position. This pressure effect on a ship's hull is created by the shape of the hull in the area of the stern. The width of the ship at the bottom of the propeller aperture is minimal, *ie* the width of the sole piece. Because of the hull shape being very fine at the bottom there is less suction behind the bottom blade. But on the top blade and above, the hull is shaping outwards vertically and horizontally, causing a greater suction effect behind the blade as it turns to the right, or to starboard when going ahead. If the right-handed propeller is moved anticlockwise into the astern position (*Fig 13.2*) the opposite suction force is effected so that the stern moves bodily to port.

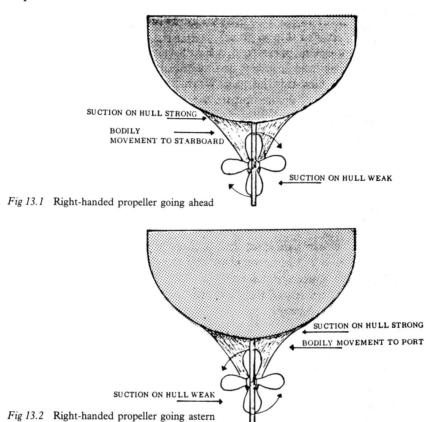

Fig 13.1 Right-handed propeller going ahead

Fig 13.2 Right-handed propeller going astern

This latter movement is the one which is important to the shiphandler. In the modern trawler the stern of the ship is full down to the water line, and below in the stern trawler. In calm conditions with still water, the propeller of a right-handed ship when put astern will draw the stern to port and the head to starboard, and we have a starboard swing. When the propeller is put ahead, the ship will have a tendency to swing to port. But if the shiphandler wishes to swing a right-handed ship in a limited space, then he must go ahead with full starboard rudder and then astern, which will not only maintain the swing but also the position of the vessel within the swinging area. Conversely, the left-handed ship will swing to port when the propeller is put astern. The shiphandler now has the following rule for swinging his ship and should, when possible:

— Swing the right-handed ship to starboard.

— Swing the left-handed ship to port.

— When going alongside, port side to, make the approach run to the berth with only a little headway, and the fore and aft line of the ship finely angled towards the quay. When closing the quay, the righthanded ship will square up with the jetty when the propeller is put astern. If the swing to starboard is too great, then, by reason of having no way on the ship after the astern movement, a touch ahead and port helm will stop the swing and keep the vessel parallel to the quay. (See *Fig 13.3a*).

— When going alongside, port side-to, in the left-handed ship, make the approach run as before. On this occasion it is even more important that headway should be minimal. When sufficiently near to the berth for a line to be put ashore from forward, put the wheel hard to starboard and, with a short turn ahead on the propeller, develop a starboard swing; stop when the swing begins. When approximately parallel to the quay, come astern so that the ship and the starboard swing come to a simultaneous stop. (*Fig 13.4c*).

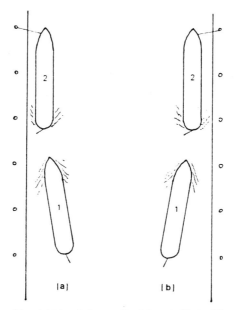

Fig 13.3 Going alongside, right-handed propeller (a) port side-to (b) starboard side-to

— When going alongside starboard side-to, in the left-handed ship, make the approach run with minimal headway towards the quay. When near enough to pass a line ashore from forward put the propeller astern. The ship's head will swing to port away from the quay and the stern will swing towards the quay. When approximately parallel to the quay put the wheel/rudder hard to starboard and with a short turn ahead on the propeller the swing will be stopped. (See *Fig 13.4c*).

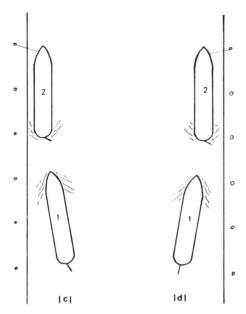

Fig 13.4 Going alongside, left-handed propeller (c) port side-to (d) starboard side-to

In all the above cases, crew should be at stations fore and aft, with heaving lines and mooring ropes ready. Breast ropes should be put out as soon as possible and the ship brought alongside easily on to fenders prior to mooring up properly shown in *Fig 13.5*.

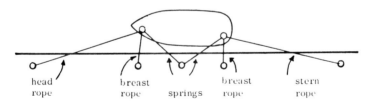

head   breast   breast   stern
rope   rope   springs   rope   rope

Fig 13.5 Correct mooring of a vessel

Effect of wind

In windy conditions the skipper should have anchors ready when going alongside. Wind, if fresh enough, may overcome transverse thrust by blowing the bow down on to a quay. If the stern sets towards a quay, the propeller and helm used appropriately will keep it off. If, however, the wind blows the bow quickly down on to the quay, then only an anchor dropped on to the bottom will save the situation. If it is known prior to berthing that the wind is strong and that there is no tug to assist, then an anchor should be used to control the bow. High sided stern trawlers, are particularly vulnerable to a lot of wind.

There are two methods of using an anchor when going to a lee berth. If in dock or mooring in shallow water, then the ship may be brought abreast of the quay at a considered safe distance. By dropping the anchor on to the bottom on a short stay, the ship's head will slowly pay off to leeward controlled by the anchor dredging along the bottom. The length of cable may be judiciously increased if movement towards the quay is too great. This method may be found useful for trawlers where they have a high raised whaleback with a much lighter forward draught than aft. The approach to the lee berth may be made at a broader angle when the anchor is used, thus keeping the stern up wind a little. If at any time too much cable is used, then the anchor may hold and bring the ship's head up into the eye of the wind. It is wise therefore to put the anchor on the bottom in good time, so that its braking effect on the ship's head can be seen and adjusted.

Anchors may be dropped and cable paid out abreast of a berth when it is known that a strong prevailing wind is likely to blow on to the quay. The anchor can then be used to heave the ship off when sailing. It is as well to know when the ship is sailing however, because an anchor laid out on the flood tide could be an embarrassment if the ship is to sail on the ebb tide.

Use of anchor when swinging

If a vessel has to be swung with a flood tide running with the ship in a narrow channel, then an anchor should always be made ready before arriving at the place where the swing is to be made. (See *Fig 13.6*) Run all the way off the ship by slowing and stopping the propeller well in advance. Go astern on arriving at the swinging area. If swinging to starboard, keep as close as possible to the port side of the fairway and have the starboard anchor ready. After having gone astern to reduce the effect of the flood tide, the skipper should not be surprised to see that the effect of transverse thrust is nil. This is because the flood tide is strong and the fore and aft direction of flood down the ship's side is far greater than the low power of the transverse thrust. With all the way off put the helm hard over to starboard and with propeller ahead bring the vessel athwart the tide. Drop the starboard anchor on to the bottom so that it just runs clear of the ship's side. Take off any surplus headway and hold on to the dredging anchor and the tide will do the rest of the work by swinging the ship on the dredged anchor, which should only be bumping and snatching at the bottom. *The important part of this operation is that of letting go the correct amount of cable.* The mate should let the anchor go when ordered to do so. When the anchor hits the bottom the cable will become slack and at this stage the brake should be applied. There are two reasons for applying the brake at this point. The cable is slack and the brake will take its weight easily. If a little more cable is required it can be paid out slowly by easing the brake. If, however, too much cable is run out, the anchor may hold on the bottom, the brake will be applied, and because the ship is athwart the tide, a tremendous weight will be taken by the windlass brake. If this happens, the brake may not hold; the cable will continue to run out and part at the end. The brake linings could burn out and the ship would swing violently.

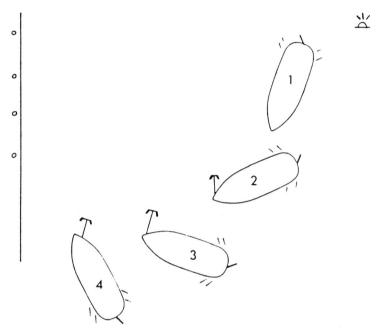

Fig 13.6 Use of anchor when swinging in a tideway

Throughout this chapter it has been stressed that a shiphandler should never have too much way on the ship when manoeuvring. This is true, and it is the basis for adept and skilful ship control. The reader should not gain the impression however that whilst handling a ship, full power either ahead or astern should not be used.

There are often occasions when bold and decisive action is necessary to avoid a hazard, when a temporary burst of full ahead, with the wheel hard over will avoid a danger. For example let us suppose that a trawler is to leave a dock and pass through a 300 ft lock, with both sets of gates open and with the dock and lock lying on an east to west direction. If we also suppose that a strong to gale force wind is blowing from the north across the lock it will be clear that on a reduced speed approach towards the lock the ship will have to be steered well up into the wind to make a good fetch at the lock. The dock being wide enough allows this to be done, but if the lockpit is only 50 ft wide, then bold action will be necessary to take the ship through the lockpit unscathed. When the ship's head is about to enter the lock, full speed will have to be used to hold the ship up to wind and steer the ship through without scraping the lee side. The full speed thrust of the propeller would give responsive and accurate steering through the lock and when clear of the lockpit the vessel would still not have gathered much headway, so that speed could be reduced again. If the beam of the ship was such that the lock was not much wider than the ship's beam, then it would be unwise to try to pass through the lock without tug assistance, and with the tugs on short bridles.

By learning and studying tides, skippers and mates leaving their home port on tidal estuaries will become proficient shiphandlers and may learn the art of letting the tide assist a manoeuvre where possible. A ship is always under better control and manageable when head to tide.

Stern to tide

When leaving a tidal jetty, head down river with the ebb away, there are dangers to be looked for. If there is plenty of room and water abreast of the jetty, the manoeuvre of letting everything go aft and holding on to a tightly set forward backspring will allow the ship's stern to cant off the jetty into the river. Let go of the spring when sufficiently well angled and at the same time put the propeller astern. The funnel effect of the ebb running down river between the quay and the ship's bow will normally keep the stern away from the jetty. Transverse thrust would most likely have little effect. When leaving a jetty on the ebb as described, too large a swing should not be allowed to develop, so that as soon as the bow is sufficiently clear from the jetty with helm and propeller ahead, the swing should be counteracted and the ship headed down river. If the ebb is strong and the stern of the ship swings quickly off into mid-channel, back off from the jetty and maintain the swing by going ahead to bring the ship head up into the ebb. If there is sufficient width, keep the vessel swinging until head down river.

If it is seen that there are ships at anchor abreast of or down river from the berth, *then under no circumstances should the above manoeuvre be carried out.* The best method under these circumstances would be to take a tug aft to hold the ship's stern up river until the vessel cleared the jetty. The tug could then be let go and the ship could proceed down river, once the tow rope was clear.

In the event that no tug was available and the ebb was not too strong, providing that there was sufficient room on the jetty, the ship might be swung on the quay. By putting out a suitably strong back spring from the fairleads at the bow, setting it tight and letting go all the after mooring ropes, the stern would swing right round for 180°, thence alongside again, leaving the ship head up river to the ebb. In small ships, such as trawlers might be considered to be, this manoeuvre is not difficult. The bow should be well fendered and the back spring sound. When this operation is carried out, it may appear that the stern will describe the 180° turn at an alarming speed, but as the stern closes the quay, there will be a cushion (pressure) effect between stern and quay. This effect, with the spring now a forward leading bow rope and the use of propeller/helm, will prevent any impact of the stern on the quay. This manoeuvre will, however, be less effective on a pile jetty.

Shallow water shiphandling

When a ship is taken into very shallow water with way on, the following effects will warn the shiphandler of the vessel's proximity to the ground:

— The bow wave and frictional wake down the side of the ship will disappear and the water will appear to be flat and calm.
— If the ship is by the stern on draught, on entering the shallow water the ship's head will visibly go down when looked at against the shore or horizon. The engines, wheelhouse, mast and superstructures will vibrate violently. The engine revolutions will drop. Only by reducing engine speed to dead slow will the vibration be stopped.
— When a vessel touches the bottom with the keel, if the bottom is of soft mud, the touch will be felt, if the bottom is hard sand, shingle, stone, *etc* it will be heard as well as felt.

It will be seen from the above that when a vessel is knowingly taken through a channel or dock approach where there is only a little more than sufficient water to float, the chance of hitting the bottom and losing control of the ship varies directly with the propeller speed plus the speed of the ship through the water.

What is actually happening is that when a ship's hull is propelled forward, the bow action pushes water away from the bows. This displacement is followed by a trough towards the after end of the ship. At full speed, even in deep water, the after draught increases, or it is said 'the ship pulls down'. In shallow water, the trough becomes deeper, because the water displaced by the hull is not so easily replaced. There is little or no water under or around the ship.

Canal effect

The pressure effects of water around a ship when steaming in a river, canal or channel, can be appreciated and will be clearly seen by watching the shore. Suppose that there is a tide gauge on the river bank, jetty or dolphin and the reading is taken well before the ship comes up to it. (This is the proper time to read a tide gauge). Further suppose that a depth of 6 m is showing on the gauge as the ship approaches, and before the bow is abeam of the gauge there is a surge of water which may well rise to about 7.5 m. As the bow passes the gauge the water level will drop quickly to about 5 m (the trough) and then rise once more (the stern wave) before settling down at 6 m.

When passing a jetty at which a ship is moored, these pressures or water surge may damage the vessel or her moorings and *that is why it is necessary for vessels to slow down before and until a jetty is passed.*

When a ship passes close to a bank and at full speed, this movement of water is restricted by the proximity of the bank and the trough becomes deeper towards the bank. Because the water on the bank side is in the form of a narrow band deepening to a trough its velocity is increased and the bow is pushed or cushioned away from the shore. The stern will be sucked towards the bank by reason of the deep trough, and the whole effect is known as interaction, canal effect.

Interaction between two ships will take place more readily when one vessel overtakes another in a fairway or channel when too close together. The overtaking ship may suck the other into her side when abeam and experience shows that a vessel on full speed passing another on slow speed will draw the overtaken vessel along for a great distance, if the channel is narrow and/or shallow.

Variable pitch propellers

The forces of a propeller when turned in water are two-fold. By far the largest component is that force which drives the ship either ahead or astern. There is a minimum component which has already been described, *ie* transverse thrust. The effect of the transverse thrust component has been explained for the traditional engine/power transmission to the propeller, where there is an engine which can be reversed in order to go astern.

A new form of transmission known as the variable pitch (VP) propeller has increased in use. The ship's engine is so arranged that it runs in one direction only. The engine revolutions may be controlled from an appropriate minimum speed to the maximum full speed. The engine is geared to the propeller shaft in such a manner that at slow speed, by means of a gear change, the direction of the shaft will be reversed (similar to the reverse gear of a car). The effect of transverse thrust at this stage is the same as that of the traditional vessel except for the fact that to the geared shaft there is a propeller fitted which has a variable pitch facility.

The VP propeller fitted on a geared shaft can be moved in one direction only, that is from a full pitch down to a zero pitch, or vice versa. Because the direction of the drive shaft can be reversed, a reverse pitch is unnecessary. Consequently a vessel wishing to go full speed will have maximum pitch on the propeller blades and maximum revolutions on the engine.

But at slow speed, not only is the engine slowed down to its minimum, the pitch may also be reduced by degrees until zero pitch is attained, when the propeller is turning as a wheel in the water, with no bias on the blades. This then is the stopped position even though the engine, shaft and propeller are turning at slow speed. The propeller is said to be feathering when turning on zero pitch.

Referring back to *Figs 13.1* and *13.2* we will see that with the VP propeller it may still be considered as righthanded or lefthanded as is the conventional propeller, depending on whether the drive shaft turns clockwise (RH) or anti-clockwise (LH) when going ahead. The suction still presses on the hull shape and follows the blade direction. But the suction diminishes as the pitch of the propeller is reduced, because the angle of the blade as a thwartship component is less, the suction behind the blade is less. When on zero pitch, however, there is still a little bias which may be described as wheeling effect. So now it can be seen that transverse thrust on a VP propeller is not quite so effective, once the shiphandler begins to reduce pitch from full. Transverse thrust either ahead or astern is affected by

— Reduction of engine speed
— Reduction of pitch.

VP propellers, reversible

There is, at the present time, in wide use within fishing fleets the fully controllable pitch propeller which is reversible. In this type of vessel both the engine and propeller shaft are uni-directional, *ie* they run continuously in one direction. The continuously running engine and shaft have speed control between slow and full speed, *ie* say 200 revs up to 600 revs.

The directional thrust of the propeller, ahead or astern, is altered by means of a reversible pitch. In short, the propeller and blades always turn in the same direction, but the pitch or angle of the blade in the vertical plane is altered so that the thrust effect in the water surrounding the propeller gives ahead propulsion or astern propulsion.

If, for example, we take a variable pitch propeller in a ship which a skipper is to join for the first time. The skipper, before handling the ship, must find out in which direction the engine (shaft) turns. Let us suppose it is clockwise. Then because the propulsion is unidirectional, the ship will have a tendency to go to port when put ahead or astern because the transverse thrust is uni-directional, *ie* the screw always turns to the right, only the blade bias changes.

In conclusion we have in the case of fully reversible pitch propellers on a uni-directional shaft and engine the following effects for shiphandlers:

— Vessel with clockwise propulsion, transverse thrust of stern to the right on ahead and astern manoeuvre, port swing.
— Vessel with anti-clockwise propulsion, transverse thrust of stern to the left on ahead and astern manoeuvre, starboard swing.

Pitch

The pitch of the screw is the distance it would move the ship ahead in one revolution, supposing that the propeller is turned in a solid and not in water. If one imagines an ordinary wood screw being turned by a screwdriver into wood, then the depth gained by one round turn is comparable to the pitch of a propeller. The greater the angle of the blades, the greater the pitch.

Slip

Slip is the difference between the actual distance covered by the ship and the pitch of the propeller/engine revs. It is mainly due to the fact that unlike a solid, water yields to the pressure exerted on it by the screw as it thrusts the ship ahead. This is important in shiphandling. If we suppose that a ship is proceeding down river on a strong ebb of five knots, then we have a situation where the water is carrying the ship at five knots regardless of engines. But because the water is moving downstream and against the after part of the propeller blades the slip will be negative or nearly so as the propeller turns against it. But what happens when, in the case of an emergency, the engines are put astern? The ebb tide continues to run past the propeller and the stern movement of the screw is in water which is passing the blades because of its own velocity. The percentage of slip becomes very high and the engines must be kept running astern for some considerable time before headway is lost.

Engine speed

Engine speed is the rate at which the propeller would drive a ship if there were to be no slip.

Rudder

The rudder in trawlers is usually placed behind the propeller so that the water thrust astern by the propeller may steer the ship. In contrast to the propeller, the greater the slip, the greater the efficiency of the rudder. To illustrate this we may look at a vessel with engines stopped and at anchor. With a strong tide running the vessel can be steered. With a strong following tide and a ship proceeding at slow speed, the rudder is less effective, there is less slip, and less water running past the rudder.

Shiphandlers should also bear in mind that in a sudden emergency, when going astern in order to get the way off, putting the wheel hard over helps the ship to pull up. The flat surface of a rudder on maximum angle being dragged astern causes some loss of headway.

Pivoting point

The average vessel will pivot approximately one third of the length of the ship measured from the bow. This is useful to know when going alongside or when swinging a single screw vessel.

Turning circle

When the helm on a vessel is put hard over she will turn on a circular path through 360°. If at sea and steaming at full speed the turning circle will be greater than that made under other circumstances, of particular interest to the shiphandler. At sea and under full speed there is a greater side slip away from the centre of the turning circle. When shiphandling, the skipper will normally be interested only in half of the turning circle, *eg* when swinging. Suppose a vessel has just picked up her anchor, is head to tide and has to turn round to proceed up or down river. The ship will have no headway so by putting the wheel hard over, propeller to full ahead, the smallest turning circle will be achieved. Because the ship begins the turn with no headway there is no side slip or moment of centrifugal force and the ship's speed over the ground will never reach maximum by the time she has swung 180°. Once again when proceeding with the tide even with propeller stopped prior to swinging, because the trawler is being carried ahead the centre of gravity of the vessel will create the centrifugal force or side slip when the turn begins. As previously suggested, have the anchor ready when turning with the tide in confined waters.

When opportunity allows, a skipper should, when at sea and on full speed make a turning circle so as to be able to judge the performance of his vessel. In fine weather and with time to spare before catching the tide, a quiet area at sea with no traffic would be a suitable time to assess the swinging of a vessel when on full speed and also from a standing start.

Use of rudder when going astern

When a vessel has gathered sternway on swinging and the swing is to continue, experience shows that by stopping the propeller and by putting the helm no more than halfway over in the direction of swing the vessel will be more likely to follow the rudder than if it were put hard over.

Going alongside a vessel at anchor

A vessel at anchor will pivot around her stem to a degree depending on wind, tide and swell. A vessel in an anchorage which is not sheltered from wind and/or swell, or one in open waters, will certainly be moving in a figure of eight motion either large or small (see *Fig 13.7*).

Fishing vessels which are called upon to go alongside a vessel at anchor should exercise the utmost caution before doing so. The first task before going alongside of the anchored vessel, after having decided that it is safe to do

128

so, is to establish communication by VHF and maintain it throughout the operation.

The approach run to the anchored vessel should not be made at a fine angle from the stern, nor should it be made with very much headway. As the bow of the approaching vessel nears the stern of the anchored vessel, even in a calm, interaction between the ships will be set up, depending on the way of the approaching vessel. The displacement effect at the bow will put pressure on the stern of the anchored vessel and push it away. The ship at anchor will pivot at an angle across the approaching vessel's bows and a collision will probably occur. If contact is made at the forepart of the anchored ship, both vessels will sheer away forward and a second contact will take place between the quarters.

It is far better to approach an anchored vessel on her beam at a suitable distance and pass breast ropes across. By doing so the ships may be brought together under the control of the breast ropes and by winch. The vessel coming alongside must in any case take action to avoid impact and if necessary steam away and try again.

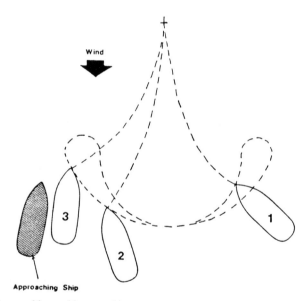

Fig 13.7 Approaching a ship at anchor

Once having moored alongside an anchored vessel it is of primary importance to be sure that the vessels are well fendered. It will also be seen that any two vessels are unlikely to have the same characteristics insofar as draught, length, beam and stability conditions. Consequently, in any sea, swell, or wind, the ships will have different roll and pitch periods, and they will range differently. In moderate to good conditions, good fendering may be adequate to prevent damage, but the prudent skipper will have the ship ready at all times so that he is ready to cast off in the event of worsening wind, sea, or swell and so avoid damage to his ship.

Shiphandling at sea

In good weather at sea shiphandling is mainly knowing how quickly the ship will react when turning, steering and stopping during an emergency. Fishing vessels have the added problem of manoeuvring with gear extended. Otherwise shiphandling at sea is primarily concerned with the safety of lives on board and the safety of the vessel in bad weather. The largest fishing vessels are about 75 m in length; most vessels are trawlers of some 60 m in length or less. So when considering bad weather, length of sea, and swell, the trawler may be thought to be a small ship relative to the prevailing weather conditions. Large vessels in a short sea behave well and are more comfortable than small vessels.

Primarily, the capacity of a trawler to ride heavy weather in a reasonable manner without risk of damage depends on the course and speed. Secondary considerations affecting good sea behaviour in heavy weather are stability range, weight distribution, and wave period. Though secondary, these conditions are very important.

Roll and pitch

There is probably little the trawlerman can do immediately insofar as altering either stability or weight distribution when heavy weather is encountered and risk of damage and seaworthiness becomes apparent. But by altering course and/or speed, the roll and pitch period will be altered. By reducing speed or altering course, the pitch and roll period will change the ship's position and direction relative to the waves.

Pounding

At the present time there are many stern trawlers of varying size which are so constructed that their engine rooms are well forward, probably only one third of the ship's length from the bow. When steaming into heavy weather these ships are particularly susceptible to pounding or slamming damage. There have been numerous cases when vessels have pushed on at full speed into a head sea when pounding occurred, and severe damage has been caused to bottom plating, twisted frames and damaged holding-down bolts on the engine. Skippers are aware that when steaming against high winds and heavy seas, water will be taken over forward causing damage on deck which will be seen. Portholes, bridge windows, bent ladders, indents to deckhouse plating are some of the deck damage which has been suffered by trawlers before the ship was eased down. The vessel should always be eased down as soon as water is being taken on board in appreciable quantities. When pounding and slamming takes place *speed should be reduced at once*. This cannot be stressed too strongly; unseen and unknown damage may be done to bottom plates and frames, especially in stern trawlers with engine rooms well forward. A stern trawler steamed north through the Minch, where she had been sheltering from bad weather from the west. When rounding the Butt of Lewis, heavy seas from the west were encountered and the vessel pounded heavily before being eased down. This modern vessel on return to port had all bottom plates under the engine room set up between frames, several frames were distorted along the bottom and up the ship's sides, three deck supporting pillars were buckled in the engine room and the engine and shaft were put out of alignment. All this damage was caused in a very short length of time when the vessel left sheltered water and turned into head seas at full speed. In case the reader should gain the impression that being head to sea is a hazardous position in bad weather, let it be said that this is not so. By design, shape, construction and stiffening, a trawler put head to sea at the right speed is likely to be much safer, less vulnerable and better to handle, than with the sea astern or on the beam. If the weather is bad enough, the trawler will be less vulnerable even if speed is reduced until only steerage way is maintained and the vessel is hove-to or is dodging. By reducing speed in heavy weather from ahead, not only does the prudent seaman avoid the risk of damage being caused forward, but also the damage which may be done to engine, gearbox and shaft which can be caused by the propeller racing, *ie* the propeller wholly or partly emerging from the sea loses the normal frictional resistance present when turning in water. As a result, propeller, shaft and engine will race at excessively high speeds.

Stiff and tender vessels

The degree to which a vessel will roll on any given wave period, when such waves are abeam, will depend upon the ship's stability. If the centre of gravity is low in the vessel there will be a large metacentric height. If the centre of gravity is high in the vessel there will be a small metacentric height. In the first case, the low centre of gravity will make the ship stiff and resistant to the waves. The roll period will be short and quick and waves will break over the ship causing damage. The quick roll will be uncomfortable, with a greater strain on lashings, stays, clips and other holding down gear due to the quick thwartship direction change.

The counterpart to the stiff ship is the ship with the high centre of gravity, which vessel is described as tender. The roll of the tender ship will be longer and slower than that of the stiff ship. If water is taken on board a tender ship, with high bulwarks and jammed freeing ports, the weight of water winged out to the sides on a long, large angled roll will dangerously reduce stability.

Both extremes of condition, stiff and tender, are undesirable. Skippers should therefore always study the stability plans on board in order to understand the ship's condition with regard to behaviour in bad weather.

Pooping and quartering seas

When a ship is steaming and the sea is on the bow, she will roll and pitch simultaneously. The vessel will encounter the waves by putting the bow or shoulder into the wave and the resistance offered reduces the angle of roll. When running with the sea on the quarter the opposite effect takes place, the roll increases because there is less resistance forward and the wave runs past the ship more slowly because ship and wave run in nearly the same direction. This combination produces a pronounced roll and pitch effect which results in heavy seas being taken on board from aft. Because the seas are moving from aft and traversing the ship, the rudder is less effective and the ship may be slewed across the wave direction and broach-to. The added weight of water taken on deck when pooped and broached-to on a large angle of heel could result in a loss of positive stability and capsizing as shown in *Figs 13.8 and 13.9.*

This situation will have arisen because of the pronounced heeling effect of the sea on the quarter; the large amount of water allowed to come on board because the ship remains under the crest of the breaking wave for a long period; or poor steering qualities.

In circumstances such as these *the skipper must take immediate action by varying both the course and the speed to avoid synchronisation of ship/wave speed and direction.* It may be decided to turn head to sea.

Turning a trawler, heavy weather

The decision to turn a trawler in heavy weather *should not be left until it is too late,* ie when a lee shore enforces the decision. Turning from a following sea until it is ahead requires skill and judgment. It should be done on a chosen time, the manoeuvre being started as soon as the last wave crest has passed the vessel. Speed should be reduced to a minimum, this will allow the last sea to pass quickly. With everything on deck and below secure, the wheel should be put hard over and the ship turned in the trough between wave crests. The trawler should be turned as quickly as possible in the trough so that she is head on or nearly head on to the next wave in the cycle. This will only be done in the lull immediately following one wave and the next. By use of full rudder the turn will be started, full speed may be used to turn short round, so long as too much headway is not gathered, before meeting the next sea. By studying the cycle of wave crests, the opportune time to turn may be chosen. Oil spread from aft prior to turning will assist greatly. Do not commit the ship to turning when water has been taken on deck. Try to turn

132

Fig 13.8 Stages in a vessel being broached-to

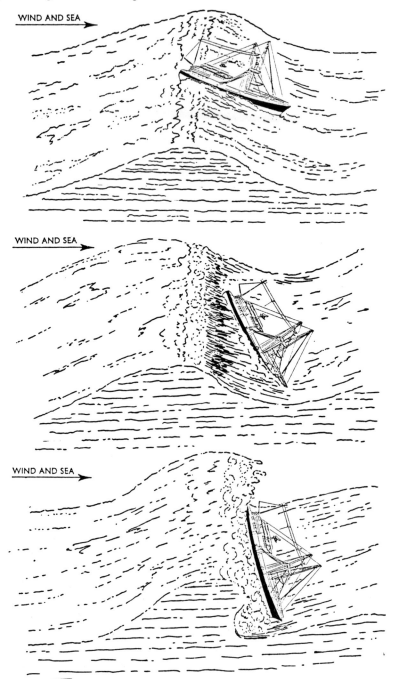

WIND AND SEA

WIND AND SEA

WIND AND SEA

Fig 13.9 Stages in a vessel being pooped

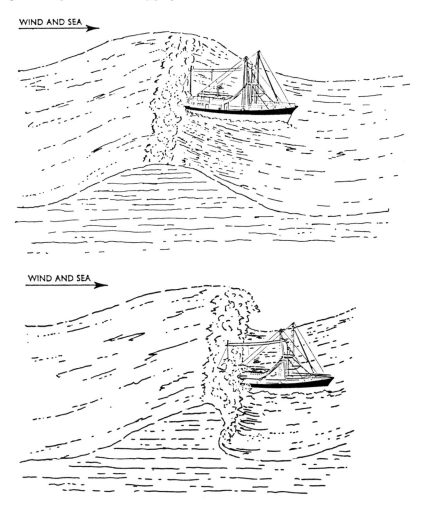

WIND AND SEA

WIND AND SEA

when the wave crest has just passed and only a comparatively small amount of water has been taken. By starting the turn at this stage on easy revolutions, the ship may, when athwart, be pushed round into wind and sea on full revolutions when in the trough. As soon as the trawler is nearly head up to wind and sea, *speed should be reduced* and the vessel steadied to meet the next sea. The prudent skipper will always turn head to sea before being forced to do so by reason of extensive damage, loss of stability due to heavy shipping of seas, or proximity of a lee shore.

A stern trawler may find that it is necessary to turn, when running before heavy seas, much earlier than the conventional side trawler, by reason of her ramp. At slow speed with a following or quartering sea, a wave will readily run up a ramp and break on to the after deck. On a heavy heel and pitch motion the water on the after deck would effect a loss of stability on the downward roll. The high bulwarks which run from right aft to the midship accommodation block on the stern trawler provide a comfortable working space in normal conditions, but in a following sea, a natural water trap exists. *Freeing ports must be kept clear and in working order and pooping avoided.*

Effect of wind on a vessel

Once a vessel has been obliged to reduce to slow speed in a storm, the pressure of the wind on her hull will have an increased effect on her handling qualities. The effect is enhanced if the ship is lightly laden, or is of shallow draught, or has large superstructures. When going very slowly or when stopped, most ships tend to lie broadside-on to the wind, and in exceptionally strong winds it may be difficult to turn them up into the wind, though it may be possible to turn them away downwind. In a typhoon or hurricane it may be impossible to turn certain ships into the wind, which is one good reason why any skipper avoids such conditions with land or dangers to leeward.

Leeway caused by the wind

The amount of leeway a ship makes in a gale depends on her speed, draught and freeboard, and on her course in relation to the direction of the wind and seas. In winds of gale or hurricane force the leeway with the wind abeam can be very considerable, and may amount to as much as two knots or more, particularly if the ship is steaming at slow speed.

It is a common mistake among inexperienced seamen to make insufficient allowance for leeway, particularly in a prolonged gale when, in addition to the wind, there will be a surface current caused by it. The amount of leeway made by a ship in various circumstances can only be judged by experience, but it is wise to allow a liberal margin of safety when passing dangers to leeward, because cases abound of ships having gone aground through failure to make sufficient allowance for leeway in the course steered.

Icing

A major danger facing fishing vessels is icing.

In certain weather conditions ice accumulating on the hull, superstructure, and rigging of a fishing vessel can be a serious danger to stability. The accumulation can occur from three causes:

— Fog, with freezing conditions, including frost 'smoke.'
— Freezing drizzle or freezing rain.

— Sea spray or sea water breaking over a ship when the air temperature is below the freezing point of sea water (about minus 2°C).

The weight of ice accumulating in rough weather with very low temperature and large amounts of spray and heavy seas breaking over a ship can be rapid. The speed with which ice accumulates increases progressively as the force of the wind increases and as the sea temperature decreases. When these conditions occur *speed must be reduced* and, if necessary, one should heave to so as to avoid heavy spray and seas breaking on board. Alternatively, seek shelter or steer towards warmer water if the direction of the wind and sea will allow this.

Safety considerations in heavy weather

Apart from his natural concern for the safety of crew working on deck, most of a skipper's responsibilities for ensuring the safety of his vessel are related to preventing unnecessary damage to it and to safeguarding its stability.

As an elementary precaution all doors and hatches through which seas can enter the hull or superstructure should be effectively closed, clips and other securing arrangements being maintained in good order to this end. Remember that intact enclosed spaces such as deckhouses contribute to the reserve buoyancy of the vessel. Plugs or other closing devices for fuel tank vents on deck should be secured. Hatch covers and deck scuttles should in any case be properly secured when not fishing.

Freeing ports should be kept clear, with any flaps fitted being free to move. Water on deck is a hazard to stability unless quickly cleared.

A fishing vessel should not be caught with its gear out at the onset of heavy weather, when a sudden deterioration in conditions may make its recovery difficult for men on deck. A 'fastener' in such circumstances could be particularly hazardous, especially if the trawl came abeam.

All fishing gear should be securely stowed and placed as low as possible; this applies particularly to beam trawlers.

Steering should be changed over from automatic to helmsman control to anticipate the effect of seas on the ship's manoeuvring, especially in quartering seas. The skipper should be alert to the need to reduce speed or alter course in response to dangerous conditions and the special care needed when deciding to alter course across a heavy sea.

14 Meteorology

Surrounding the earth and revolving with it through space we have an atmosphere which lies in the form of an envelope through which we can see other heavenly bodies. The layer of atmosphere nearest to the earth is known as the troposphere and has an average depth of from 5 to 10 miles.

Within the troposphere there are air pressure and temperature variations. There is heat radiation from the sun as well as heat diffused by the earth itself. The earth revolves on its axis and we have large areas of ocean with irregularly shaped land masses. These factors all contribute to meteorological conditions and approximate behaviour patterns of weather systems.

The phenomena we describe as weather are caused by movements and changing conditions of the troposphere, containing principally oxygen, nitrogen and varying amounts of water vapour which cause cloud, rain and fog and make for equable temperature. Changes in atmospheric pressure, commonly measured in millibars, are associated with changes of weather, especially of the wind. Broadly speaking high pressure is associated with fair, though not necessarily sunny, weather and light winds, while low pressure usually entails strong winds and rain. Wind is the result of a difference of pressure between two adjacent areas.

Temperature

The heat of the sun is a primary cause of motion of the atmosphere; hence weather is closely bound up with changes of temperature. The sea requires more heat than land to raise its temperature and yields more heat than land with a fall in temperature.

Humidity

The amount of water vapour in the air varies and is an important factor in determining the nature of the weather. The amount of water vapour that the air can hold in an invisible state is limited and varies with the temperature. Warm air can hold more water vapour than cold; when the air contains as much water vapour as it can hold at a given temperature it is said to be 'saturated'. The temperature at which a mass of air reaches saturation is called the 'dew-point'. If the temperature is lowered or more water vapour is added to the air, the surplus condenses into minute but visible drops of water, which constitute cloud, mist or fog.

Humidity is measured with a hygrometer, the commonest form of which is the 'wet and dry bulb thermometer', described on page 72. When the air is dry the wet bulb reading is several degrees below that of the dry bulb, but when it is saturated both will read the same and fog may then be expected.

Cloud

We see then that when air is cooled below its saturation point the surplus water vapour condenses into visible droplets of water. Such condensation occurs in the atmosphere when there is cooling due to uplift and results in the formation of cloud. It is largely upon the manner of uplift that the type of cloud depends.

Air is often heated in its lower layers by its passage over a relatively warm land or sea surface. This causes it to expand, become lighter and rise. As the air rises it expands further because of the decrease in pressure, and cools because of its expansion. If this cooling is continued beyond the point at which the air is saturated, cloud results. Clouds formed in this way are of the *cumulus* type, with horizontal bases and rounded heads which sometimes spread out at great heights into the shape of an anvil.

Where two masses of air from different sources and with different characteristics meet, boundary lines are formed where the colder and heavier air runs under the warmer and lighter air and compels it to rise. As in the previous case, the expansion and cooling of this air results in the formation of cloud, which spreads as a more or less continuous layer of *stratus* type cloud.

Rain, snow and hail

If the cooling process which gives rise to cloud-formation is continued, the droplets of which the cloud consists grow in size until they become too heavy to remain suspended in the air, and they therefore fall to the ground as very fine rain or drizzle. If the droplets are held aloft by upward currents of air they grow by agglomeration and eventually fall as raindrops. When very strong up-draughts exist, as in large *cumulus* type clouds, the drops become very large before falling as heavy showers of rain or, under certain conditions, as hail. The latter is caused by very strong up-draughts carrying the rain-drops above the freezing-level, where they turn into growing balls of ice which eventually fall as hailstones. Thunderstorms are often accompanied by hail, because they also are caused by very strong up-draughts extending to great heights.

If the temperature at which condensation actually takes place is below freezing-point the droplets form as ice crystals, which grow and become snowflakes. (Snowflakes often melt and change to raindrops before reaching the ground).

Fog and mist

The cause of fog and mist is fundamentally the same as that of cloud, *ie* the cooling of air below the temperature at which it is saturated. In cloud the resulting condensation into visible particles takes place aloft, but in fog or mist the condensation occurs on the surface of the sea (or ground.) The difference between fog and mist is merely one of degree.

In the open sea almost all fog is caused by warm, moist air blowing over a relatively cold sea surface, whereby the lower layers of the air are chilled. These sea fogs are therefore most common in spring and early summer. They are also particularly prevalent in the vicinity of cold ocean currents.

Over the land, on clear nights with little wind, a sharp drop in temperature occurs after sunset as a result of the Earth radiating its heat away into space. The air in contact with the ground is cooled, condensation takes place and fog is formed. Such land-formed fogs are most common in winter. They may drift seaward under the influence of offshore winds, thus giving rise to coastal fog. If the sky is cloudy, fogs of this type are improbable, because the clouds largely prevent the Earth losing heat by radiation.

As fog can only occur when the air is virtually saturated, the hygrometer (wet and dry bulb thermometers) provides a useful guide to its probability.

Pressure systems

The barometric pressure, the temperature, and weather forecasts which are broadcast regularly are the aids by which the fisherman may preserve the safety of his ship and crew. The distribution of air pressure over the globe reveals that the atmospheric pressure around the earth's surface is roughly divided into about five main sea areas. Over ocean tropical areas we have moderately low pressure areas to the north and south of the equator. To the north and south of the equatorial zones in latitudes $20° - 40°$ we have high pressure areas, *ie* in the Pacific, Atlantic and Indian Ocean.

Higher latitudes are characterized by a series of low pressure systems or depressions which generally travel in an easterly direction across the oceans. In the North Atlantic, for example, the high pressure area (Azores) extends from approximately the latitude of the Straits of Gibraltar down to the beginning of the tropical zone at $22°N$, with the centre lying towards the eastern half of the ocean. The winds of the system, because of the earth's rotation, revolve around the high in a clockwise direction. In the south Atlantic winds will revolve in an anti-clockwise direction. The first rule of meteorology may now be set out as:

North hemisphere	Anticyclone (high). Wind revolves clockwise.
	Cyclone (low). Winds revolve anti-clockwise.
South hemisphere	Anticyclone (high). Wind revolves anti-clockwise.
	Cyclone (low). Winds revolve clockwise.

Referring once again to the anticyclone areas in the North Atlantic, it will be seen that the prevailing winds around this centre are the northeast trade winds blowing towards the tropics on the east of the system, Easterly winds on the south of the system, southerly winds on the USA side of the system, and finally southwest to westerly winds to the north of the system blowing towards Europe. These are the prevailing winds.

The influence of the sun

The influence of the sun is such that the centre of the anticyclone varies with its declination. The calm belt of the tropical zone is the developing ground for tropical revolving storms which form where the sea temperature is high causing rapid evaporation of water vapour. The energy is released in one of these storms as this vapour condenses and an area of low pressure develops, producing extremely strong winds, high seas and torrential rainfall.

Meteorological terms

The following meteorological terms are set out below in order that the figures and explanation may be fully understood.

Isobar. A line joining places of equal barometric pressure. The closer the isobars are drawn together the greater the rise or fall in the atmospheric pressure in that area, and therefore, the stronger the winds.

Pressure gradient. The difference in pressure in unit distance measured at right angles to the isobars.

Path. The expected course or direction which the storm might take.

Track. The course over which the storm centre has already passed.

Indraught. The angle between the isobars and wind direction, always towards the centre of the low.

Vortex or Eye. The storm centre, usually an area of temporary calm.

Dangerous semi-circle. The half of the storm which lies to the right of the path in the northern hemisphere. (To the left in the southern hemisphere).

Dangerous quadrant. The leading quadrant of the dangerous semi-circle wherein the wind blows towards the path.

140

Navigable semi-circle. The half of the storm field which lies to the left of the path in the northern hemisphere. (To the right in the southern hemisphere).

Trough line. The line drawn through the eye of the storm at right angles to the path.

Veer. A wind is said to veer when it changes direction in a clockwise movement from whence it blows, *ie* from south to west.

Back. A wind is said to back when it changes direction in an anti-clockwise movement from whence it blows, *ie* from south to east.

Tropical revolving storms

In the North Atlantic the path of a tropical revolving storm (hurricane) follows the line of least resistance by moving in a westerly direction below the North Atlantic high, curves on the approach to the American land mass towards the north, and then moves in an east to northeast direction towards Europe. A tropical revolving storm forming in the tropical zone will at first intensify and move at varying speeds, the average speed near to the West Indies being about 300 miles per day. On its approach to, or when crossing land, it tends to fill and dissipate because the supply of warm moist air necessary to its existence is no longer present.

Occasionally a tropical revolving storm will continue into temperate latitudes, gradually changing its character and becoming a depression.

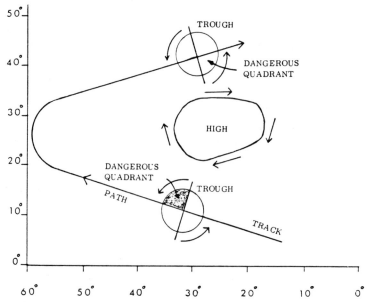

Fig 14.1 Path of tropical revolving storm in the North Atlantic

The winds circulating around a low pressure area blow slightly inwards towards the centre at an angle which is known as the angle of indraught. It will also be seen from *Fig 14.2* why in an intense low *it is hazardous to be caught in the dangerous quadrant.*

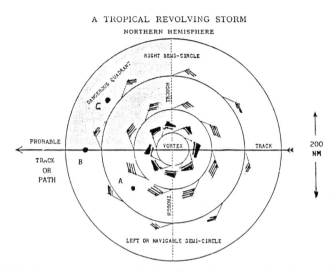

Fig 14.2 Tropical revolving storm in Northern Hemisphere giving three ships positions relative to storm centre

The position of the storm centre relative to the ship's position must be established when it is believed to be approaching. In order to take a bearing of the storm centre, we have to use Buys Ballot's Law which says that if an observer faces the true wind, then he will have the eye of the storm centre approximately 10 points to the right in the Northern hemisphere. (To the left in the Southern hemisphere). This is one of the most reliable rules in meteorology. To obtain this bearing, the vessel should be stopped in order to assess the direction of the true wind. If and when the wind changes and a second bearing were to be taken, whilst the distance away from the centre would not be known, an indication of its approximate path would be obtained.

If we look at *Fig 14.2* we will see the letters A, B, C shown as observers on the leading edge of a storm field:

A   By facing the westerly wind the observer would know that the storm centre lies to his right. If the wind backs to the left, the observer would know that he was in the left hand or navigable semi-circle. The action to take in this case would be to run with the wind on the starboard quarter until the barometer began to rise and it was confidently felt that the storm centre had passed.

B   If the wind at position B remained constant by direction, but increased steadily in force, observer B would know that he would be on the path of the storm centre. If the vessel remains here she will be overtaken by the storm centre and her experience can be imagined.

The action to take is similar to that taken by A: run to the south with the wind on the starboard quarter until the barometer begins to rise and until confident that the vortex is past and clear.

C   Being in the dangerous quadrant, the observer should steam his vessel away from the path in a northerly direction as soon as he is aware of his position relative to the storm centre. If left until too late, high seas and swell may force the ship to heave to and fall back towards the passing storm centre. A falling barometer, rising temperature and veering wind will show that C is in the dangerous quadrant.

Southern hemisphere, tropical storms

Whereas in the northern hemisphere the winds revolve around the centre of a tropical storm or low pressure area in an anti-clockwise direction, in the southern hemisphere the winds revolve around the storm centre in a clockwise direction. As previously stated the breeding ground for tropical storms is in an area which is but a few degrees north or south of the equator, usually where there is a large expanse of ocean.

The track followed by tropical storms in both hemispheres is towards the west and away from the equator until a latitude of 20° to 30° north or south is reached when they recurve towards the east, and polewards. In the South Pacific the cyclone will usually generate to the Northeast of the Fiji Islands, then follow a WSW path until latitude 20°S is reached before recurving to follow a southeast path.

Because of the clockwise wind circulation the dangerous semi-circle of the cyclone in the southern hemisphere is the left hand quadrant. The vortex of the storm may be estimated by facing the wind and measuring approximately 8 to 10 points to the *left*. This will indicate the direction of the storm centre. See *Fig 14.3* which shows the path of a tropical storm in the southern hemisphere and application of Buys Ballot's Law for the southern hemisphere.

By looking at *Fig 14.4* and supposing that three ships are at positions A, B & C it will be seen that they will find their position relative to the storm centre by firstly noting a rapidly falling barometer, increased seas and stronger winds.

A   By facing the wind when hove-to ship A will note that the storm centre is approximately 90°–110° to the left, because the wind is blowing from the southeast. If the wind backs with a continuing fall in pressure then A must assume that she is in the dangerous quadrant and steam to the southward with the wind on the port quarter.

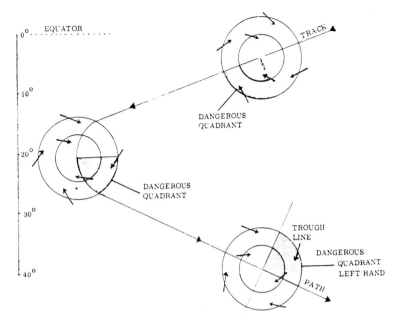

Fig 14.3 Path of a tropical revolving storm in the southern hemisphere

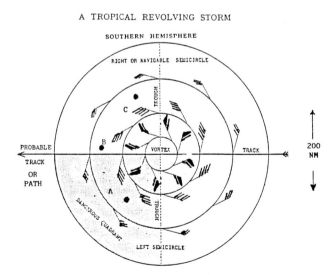

Fig 14.4 Tropical revolving storm in Southern Hemisphere giving three ships positions relative to storm centre

144

B   Ship B hove-to and facing the wind will find the barometer falling very rapidly and if the wind remains constant in direction from the south but increasing in force then the storm centre in relation to the rule must remain constant relative to the ship's position. B must assume that she is near to or on the path of the storm and should steam in a northerly direction into the navigable semi circle, keeping the wind on the port quarter so as not to be drawn by the angle of indraught towards the vortex.

C   Ship C will find the barometer falling quickly with increased wind speed from a SSW direction changing to SW and will know by using Buys Ballot's Law that the storm centre will pass south of her position. She may run with the wind on the port quarter away from the storm centre or if circumstances allow, heave to until the trough line passes and the barometer begins to rise again. The wind will veer through West and thence North.

Depressions or lows

In the higher latitudes temperate depressions or 'lows' form as the result of interaction between relatively cool dense air on the one hand and warm light air on the other. The boundary between two such air streams is known as a polar front because the cold air has a tendency to push its way forward in the form of a wedge under the warm air so that a bulge develops. A depression is formed with a frontal system which is preceded by a warm front closely followed by the cold front with the tip of the front being at the centre of the low.

Those fishermen who operate mainly in the northwest Atlantic zone will already know that depressions or lows in the majority of cases move in a west to east direction across the north Atlantic towards northwest Europe.

The wind around a low in the northern hemisphere circulates in an anti-clockwise direction. In the southern hemisphere the reverse applies.

Temperate storms

Fishermen having made themselves familiar with the circulation of winds around a storm centre must also be able to recognise a frontal system and the approach of a frontal system. The obvious and easiest way in which to learn of the existence, position and path of a depression relative to one's own position is by listening carefully to the weather forecast and synopsis broadcasts, and by listening to coastal stations which broadcast weather reports in plain language.

If, however, the fisherman has to determine whether a depression is approaching, then the barometer and thermometer are his best aids when used together. Even if it known that a low is approaching, one *can never be sure of* its path.

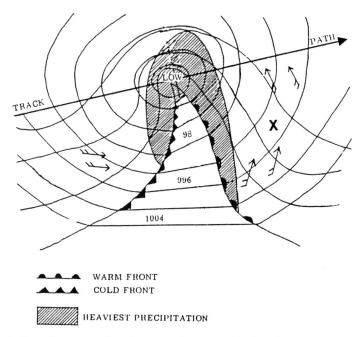

WARM FRONT
COLD FRONT
HEAVIEST PRECIPITATION

Fig 14.5   Typical temperate frontal system of a storm – Northern hemisphere

If we look at *Fig 14.5* and suppose that the fishing vessel is at position 'x' in the path of a storm moving from west to east, the pressure will fall slowly as the temperature rises, the wind will veer from south to southwest, the sky will become overcast with drizzle and a swell will become noticeable. The clouds will be high and possibly broken. The wind will continue to veer, the pressure will fall steadily, and there will be a continuing rise in temperature and heavier rain. Medium to low cloud and strong winds from the west will indicate the approach of the warm front. The time taken for these changes depends on the pressure gradient and distance away from the centre of the low.

In the warm sector between the warm and cold fronts, the barometer will remain steady as will the thermometer. The wind will be steady with poor visibility in drizzle or light rain.

On the cold front the first indications will be a sudden drop in temperature with a rising barometer, a sudden veer of the wind towards the northwest in squalls, and heavy rain or sleet with an improvement in the visibility as the rain turns to showers when the front has passed. Thereafter the weather will improve with the barometer and thermometer steadying, but with the sea and swell remaining.

Not all depressions are likely to be dangerous. Depressions or lows are a common occurrence in and around the British Isles for example, and in many cases, the centre of a depression may have a barometric pressure of 1004 or 1008 in a large field, with the isobars well spread (low gradient). At the centre, winds will probably be less than 30 knots, force 6 possibly force 7. They are known as shallow depressions but should be carefully watched in case they deepen.

It is the gale warning for gale force, storm and hurricane strength winds against which precautions should be taken. Fishing vessels should not be caught in the dangerous quadrant. In high latitudes in winter the passing of a frontal system in a storm becomes dangerous for two reasons. Firstly because it is a storm or intense depression the isobars will be close and there will be high winds and rough seas. The vessel will possibly ship water and a lot of spray. Secondly, when the cold front passes the ship's position, there is a sudden and appreciable drop in air temperature. If the air temperature drops to $-2°$ Centigrade $(28\frac{1}{2}°F)$, salt water when airborne will freeze. Consequently masts, derricks, stays, deckhouses and winches will become covered with ice. Conditions in the storm will be such that it will not be possible for men to go out on deck in order to remove the ice. The ship will become overloaded with the top weight of ice and may capsize.

The prudent fisherman will listen carefully to radio gale warnings and in the event of a storm warning, especially in high latitudes in winter, will watch the barometer, thermometer and the wind to avoid being caught in the dangerous quadrant and associated frontal system, which will give rise to freezing spray, rain and sleet.

Prevailing winds – eastern Atlantic

In the eastern Atlantic the majority of low pressure systems pass to the north of the British Isles, subjecting the area to a fair amount of southwest to westerly winds. The associated fronts of the low pressure system passing to the north from west to east are a predominant feature of British coastal weather. Fishermen who have experience of working in the western approaches to the English channel, for example, will know that southerly winds freshening, with an accompanying swell, indicate the approach of bad weather with the wind veering to the southwest to west and then northwest when the low has passed, thence moderating and leaving the swell.

Anticyclones

Anticyclones or high pressure areas are mainly considered to be good weather systems. The air around the centre circulates in a clockwise direction in the northern hemisphere (the reverse applies in the southern hemisphere).

Anticyclones are associated with small pressure gradients and light winds. They may be formed by either cold or warm air gathering over a sea or land mass. The warm anticyclone forms when the air in the troposphere is warmer than the surrounding air, the pressure being caused by the great vertical height of the warm air. An example of these types are the high pressure areas, previously mentioned, in the North Atlantic with their source near to the Azores. This is known as a maritime tropical anticyclone. In the same latitude there is the continental tropical anticyclone with its source in the Sahara. Towards the pole we have cold anticyclones which form because high pressure is created when the air is colder in one particular area compared to the cold surrounding air. Typical example of these cold anticyclone systems is the maritime polar of Greenland and the Arctic. The continental polar is the well known high in Siberia. Because the air in permanent areas of high pressure, *eg* Greenland and the Arctic, is unstable, depressions may form, as mentioned earlier in this chapter. The very cold air at the centre of the high will diverge at the base by pushing out into the relatively warmer air surrounding the high and a depression will form.

Anticyclones are good weather systems with very light winds. However, they have one drawback – fog. Fog may form when warm air from a warm high system passes over the sea which has a lower temperature, conduction takes place and fog forms.

With a very cold anticyclone the cold air meeting the relatively warmer Arctic water will create fog or Arctic smoke. This occurs in places other than the Arctic and can actually be seen to happen in rivers and estuaries in the UK. There is insufficient wind to disperse fog in anticyclonic conditions, but the sun may absorb the fog by its heat. In winter, however, there is usually little heat from the sun and the fog may well persist for long periods.

During summer in high pressure systems when there is little or no cloud cover, the land which has been heated during the day by the sun will radiate this heat at night when the air has cooled, and fog will form during the night. Sea fog will develop in summer where warm tropical maritime air crosses cold sea surfaces; the air is cooled below its dew point and fog forms.

Record of weather conditions

It is imperative to keep a continuous record of weather conditions in the bridge log book. Wind direction, force, barometer readings and temperature should be recorded accurately and regularly so that they can be referred to and used at any time when a storm is expected.

Tables of weather conditions are shown on the Beaufort scale of wind force.

148

Beaufort scale of wind force

Beaufort force	Limits of wind speed (knots)	Descriptive term	Sea criterion	Probable height of waves (metres*)	Probable maximum height of waves (metres*)
0	Less than 1	Calm	Sea like a mirror.	—	—
1	1-3	Light air	Ripples with the appearance of scales are formed, but without foam crests.	0·1	0·1
2	4-6	Light breeze	Small wavelets, still short but more pronounced, crests have glassy appearance and do not break.	0·2	0·3
3	7-10	Gentle breeze	Large wavelets. Crests begin to break. Foam of glassy appearance. Perhaps scattered white horses.	0·6	1·0
4	11-16	Moderate breeze	Small waves, becoming longer; fairly frequent white horses.	1·0	1·5
5	17-21	Fresh breeze	Moderate waves, taking a more pronounced long form; many white horses are formed. (Chance of some spray.)	2·0	2·5
6	22-27	Strong breeze	Large waves begin to form; the white foam crests are more extensive everywhere (Probably some spray.)	3·0	4·0
7	28-33	Near gale	Sea heaps up and white foam from breaking waves begins to be blown in streaks along the direction of the wind.	4·0	5·5

(continued)

Beaufort force	Limits of wind speed (knots)	Descriptive term	Sea criterion	Probable height of waves (metres★)	Probable maximum height of waves (metres★)
8	34-40	Gale	Moderately high waves of greater length; edges of crests begin to break into spindrift. The foam is blown in well-marked streaks along the direction of the wind.	5·5	7·5
9	41-47	Strong gale	High waves. Dense streaks of foam along the direction of the wind. Crests of waves begin to topple, tumble and roll over. Spray may affect visibility.	7·0	10·0
10	48-55	Storm	Very high waves with long overhanging crests. The resulting foam in great patches is blown in dense white streaks along the direction of wind. On the whole the surface of sea takes a white appearance. Tumbling of the sea becomes heavy and shock-like. Visibility affected.	9·0	12·5
11	56-63	Violet storm	Exceptionally high waves. (Small and medium-sized ships might be for a time lost to view behind the waves.) The sea is completely covered with long white patches of foam lying along the direction of the wind. Everywhere the edges of the wave crests are blown into froth. Visibility affected.	11·5	16·0
12	64 and over	Hurricane	The air is filled with foam and spray. Sea completely white with driving spray; visibility very seriously affected.	14 or over	—

★These columns are added as a guide to show roughly what may be expected in the open sea, remote from land. In enclosed waters, or when near land with an off-shore wind, wave heights will be smaller, and the waves steeper.

State of sea

Appearance	Height (approx) in metres
Calm – glassy	0
Calm – rippled	0 – 0·1
Smooth – wavelets	0·1 – 0·5
Slight	0·5 – 1·25
Moderate	1·25 – 2·5
Rough	2·5 – 4
Very rough	4 – 6
High	6 – 9
Very high	9 – 14
Phenomenal	Over 14

State of swell

Length	Height (approx)	
	None	Low
0–90 m	Short	
over 183 m	Long	about 2 m
0–90 m	Short	Moderate height
90–183 m	Average	
over 183 m	Long	2–4 m
0–90 m	Short	heavy
90–183 m	Average	
over 183 m	Long	over 4 m
	Confused	

Visibility Scale

Dense fog	Less than 50 metres.
Thick fog	50–200 metres.
Fog } *	200–500 metres.
Moderate fog	500–1,000 metres (approx. 500 m–⅝ n.miles).
Mist or haze	1–2 km (approx. ⅝–1 n.miles).
Poor visibility	2–4 km (approx. 1–2 n.miles).
Moderate visibility	4–10 km (approx. 2–6 n.miles)
Good visibility	10–20 km (approx. 6–12 n.miles).
Very good visibility } †	20–50 km (approx. 12–30 n.miles).
Excellent visibility	50 km or more (30 n.miles or more).

*Grouped together as 'fog'
†Grouped together as 'good visibility'

Beaufort notation to indicate the state of the weather

Weather	Beaufort letter
Blue sky (0 – 2/8 clouded)	b
Sky partly clouded (3 – 5/8) clouded)	bc
Cloudy (6 – 8/8 clouded)	c
Drizzle	d
Wet air (without precipitation)	e
Fog	f
Gale	g
Hail	h
Precipitation in sight of ship or station	jp
Line squall	kq
Storm of drifting snow	ks
Sandstorm or duststorm	kz
Lightning	l
Mist	m
Overcast sky. (The whole sky covered with unbroken cloud)	o
Squally weather	q
Rain	r
Sleet (i.e. rain and snow together)	rs
Snow	s
Thunder	t
Thunderstorm with rain or snow	tlr or tls
Ugly threatening sky	u
Unusual visibility	v
Dew	w
Hoar frost	x
Dry air	y
Dust haze	z

15   Collision avoidance regulations

Experience has shown over many years that numerous mishaps and incidents involving the loss of lives and vessels have been caused or contributed to by bad lookout, negligent navigation and poor bridge discipline. All fishermen should make every effort to avoid dangerous situations by adhering to the Collision Regulations and by keeping a good lookout, by navigating properly and by maintaining good bridge disciplines.

Traffic separation schemes (Rule 10)

This rule must be clearly understood by those engaged in fishing in an area where there is a traffic separation scheme, see page 157. Vessels engaged in fishing or on passage in a separation scheme are considered to be using the scheme and must conform to the essential principles of Rule 10b and c, *ie* proceed in the appropriate lane in the general direction of the traffic flow for that lane; so far as practicable keep clear of a separation zone or line; normally join or leave a traffic lane at the termination of the lane, but when joining or leaving at the side, vessels should do so at as small an angle to the general direction of the traffic flow as possible.

Fishing vessels may *engage in fishing within a separation zone* and they may follow any course *within the zone*.

Fishing vessels *fishing within a lane* should not impede through traffic. This means that they should not operate in such a manner that they or their gear seriously restrict the sea room available to other vessels within the lane. They should make every endeavour whilst fishing to avoid interfering with traffic; but nonetheless, if risk of collision with another ship develops, then the normal steering and sailing rules apply.

Skippers should avoid crossing lanes if possible but if obliged to do so they should cross as near as practicable at right angles to the general direction of traffic flow. Under no circumstances should a fishing vessel, whether fishing or not, proceed against the traffic flow in the wrong lane. *This is a punishable offence.*

INTERNATIONAL REGULATIONS FOR PREVENTING COLLISIONS AT SEA, 1972

(as amended by Resolution A464(XII))

PART A. GENERAL

RULE 1

Application

(*a*) These Rules shall apply to all vessels upon the high seas and in all waters connected therewith navigable by seagoing vessels.

(*b*) Nothing in these Rules shall interfere with the operation of special rules made by an appropriate authority for roadsteads, harbours, rivers, lakes or inland waterways connected with the high seas and navigable by seagoing vessels. Such special rules shall conform as closely as possible to these Rules.

(*c*) Nothing in these Rules shall interfere with the operation of any special rules made by the Government of any State with respect to additional station or signal lights, shapes or whistle signals for ships of war and vessels proceeding under convoy, or with respect to additional station or signal lights or shapes for fishing vessels engaged in fishing as a fleet. These additional station or signal lights, shapes or whistle signals shall, so far as possible, be such that they cannot be mistaken for any light, shape or signal authorized elsewhere under these Rules.

(*d*) Traffic separation schemes may be adopted by the Organization for the purpose of these Rules.

(*e*) Whenever the Government concerned shall have determined that a vessel of special construction or purpose cannot comply fully with the provisions of any of these Rules with respect to the number, position, range or arc of visibility of lights or shapes, as well as to the disposition and characteristics of sound-signalling appliances, without interfering with the special function of the vessel, such vessel shall comply with such other provisions in regard to the number, position, range or arc of visibility of lights or shapes, as well as the disposition and characteristics of sound-signalling appliances, as her Government shall have determined to be the closest possible compliance with these Rules in respect of that vessel.

RULE 2

Responsibility

(*a*) Nothing in these Rules shall exonerate any vessel, or the owner, master or crew thereof, from the consequences of any neglect to comply with these Rules or of the neglect of any precaution which may be required by the ordinary practice of seamen, or by the special circumstances of the case.

(*b*) In construing and complying with these Rules due regard shall be had to all dangers of navigation and collision and to any special circumstances, including the limitations of the vessels involved, which may make a departure from these Rules necessary to avoid immediate danger.

RULE 3

General definitions

For the purpose of these Rules, except where the context otherwise requires:

(*a*) The word "vessel" includes every description of water craft, including non-displacement craft and seaplanes, used or capable of being used as a means of transportation on water.

(*b*) The term "power-driven vessel" means any vessel propelled by machinery.

(*c*) The term "sailing vessel" means any vessel under sail provided that propelling machinery, if fitted, is not being used.

(*d*) The term "vessel engaged in fishing" means any vessel fishing with nets, lines, trawls or other fishing apparatus which restrict manoeuvrability, but does not include a vessel fishing with trolling lines or other fishing apparatus which do not restrict manoeuvrability.

(*e*) The word "seaplane" includes any aircraft designed to manoeuvre on the water.

(*f*) The term "vessel not under command" means a vessel which through some exceptional circumstance is unable to manoeuvre as required by these Rules and is therefore unable to keep out of the way of another vessel.

(*g*) The term "vessel restricted in her ability to manoeuvre" means a vessel which from the nature of her work is restricted in her ability to manoeuvre as required by these Rules and is therefore unable to keep out of the way of another vessel. The term "vessels restricted in their ability to manoeuvre" shall include but not be limited to:

    (i) a vessel engaged in laying, servicing or picking up a navigation mark, submarine cable or pipeline;

    (ii) a vessel engaged in dredging, surveying or underwater operations;

    (iii) a vessel engaged in replenishment or transferring persons, provisions or cargo while underway;

    (iv) a vessel engaged in the launching or recovery of aircraft;

    (v) a vessel engaged in mineclearance operations;

    (vi) a vessel engaged in a towing operation such as severely restricts the towing vessel and her tow in their ability to deviate from their course.

(*h*) The term "vessel constrained by her draught" means a power-driven vessel which because of her draught in relation to the available depth of water is severely restricted in her ability to deviate from the course she is following.

(*i*) The word "underway" means that a vessel is not at anchor, or made fast to the shore, or aground.

(*j*) The words "length" and "breadth" of a vessel mean her length overall and greatest breadth.

(*k*) Vessels shall be deemed to be in sight of one another only when one can be observed visually from the other.

(*l*) The term "restricted visibility" means any condition in which visibility is restricted by fog, mist, falling snow, heavy rainstorms, sandstorms or any other similar causes.

PART B. STEERING AND SAILING RULES

Section I. Conduct of vessels in any condition of visibility

RULE 4

Application

Rules in this Section apply in any condition of visibility.

RULE 5

Look-out

Every vessel shall at all times maintain a proper look-out by sight and hearing as well as by all available means appropriate in the prevailing circumstances and conditions so as to make a full appraisal of the situation and of the risk of collision.

RULE 6

Safe speed

Every vessel shall at all times proceed at a safe speed so that she can take proper and effective action to avoid collision and be stopped within a distance appropriate to the prevailing circumstances and conditions.

In determining a safe speed the following factors shall be among those taken into account:

(a) By all vessels:
   (i) the state of visibility;
   (ii) the traffic density including concentrations of fishing vessels or any other vessels;
   (iii) the manoeuvrability of the vessel with special reference to stopping distance and turning ability in the prevailing conditions;
   (iv) at night the presence of background light such as from shore lights or from back scatter of her own lights;
   (v) the state of wind, sea and current, and the proximity of navigational hazards;
   (vi) the draught in relation to the available depth of water.

(b) Additionally, by vessels with operational radar:
   (i) the characteristics, efficiency and limitations of the radar equipment;
   (ii) any constraints imposed by the radar range scale in use;
   (iii) the effect on radar detection of the sea state, weather and other sources of interference;
   (iv) the possibility that small vessels, ice and other floating objects may not be detected by radar at an adequate range;
   (v) the number, location and movement of vessels detected by radar;
   (vi) the more exact assessment of the visibility that may be possible when radar is used to determine the range of vessels or other objects in the vicinity.

RULE 7

Risk of collision

(*a*) Every vessel shall use all available means appropriate to the prevailing circumstances and conditions to determine if risk of collision exists. If there is any doubt such risk shall be deemed to exist.

(*b*) Proper use shall be made of radar equipment if fitted and operational, including long-range scanning to obtain early warning of risk of collision and radar plotting or equivalent systematic observation of detected objects.

(*c*) Assumptions shall not be made on the basis of scanty information, especially scanty radar information.

(*d*) In determining if risk of collision exists the following considerations shall be among those taken into account:

    (i) such risk shall be deemed to exist if the compass bearing of an approaching vessel does not appreciably change;

    (ii) such risk may sometimes exist even when an appreciable bearing change is evident, particularly when approaching a very large vessel or a tow or when approaching a vessel at close range.

RULE 8

Action to avoid collision

(*a*) Any action taken to avoid collision shall, if the circumstances of the case admit, be positive, made in ample time and with due regard to the observance of good seamanship.

(*b*) Any alteration of course and/or speed to avoid collision shall, if the circumstance of the case admit, be large enough to be readily apparent to another vessel observing visually or by radar; a succession of small alterations of course and/or speed should be avoided.

(*c*) If there is sufficient sea room, alteration of course alone may be the most effective action to avoid a close-quarters situation provided that it is made in good time, is substantial and does not result in another close-quarters situation.

(*d*) Action taken to avoid collision with another vessel shall be such as to result in passing at a safe distance. The effectiveness of the action shall be carefully checked until the other vessel is finally past and clear.

(*e*) If necessary to avoid collision or allow more time to assess the situation, a vessel shall slacken her speed or take all way off by stopping or reversing her means of propulsion.

RULE 9

Narrow channels

(*a*) A vessel proceeding along the course of a narrow channel or fairway shall keep as near to the outer limit of the channel or fairway which lies on her starboard side as is safe and practicable.

(*b*) A vessel of less than 20 metres in length or a sailing vessel shall not impede the passage of a vessel which can safely navigate only within a narrow channel or fairway.

(*c*) A vessel engaged in fishing shall not impede the passage of any other vessel navigating within a narrow channel or fairway.

(*d*) A vessel shall not cross a narrow channel or fairway if such crossing impedes the passage of a vessel which can safely navigate only within such channel or fairway. The latter vessel may use the sound signal prescribed in Rule 34(*d*) if in doubt as to the intention of the crossing vessel.

(*e*) (i) In a narrow channel or fairway when overtaking can take place only if the vessel to be overtaken has to take action to permit safe passing, the vessel intending to overtake shall indicate her intention by sounding the appropriate signal prescribed in Rule 34(*c*)(i). The vessel to be overtaken shall, if in agreement, sound the appropriate signal prescribed in Rule 34(*c*)(ii) and take steps to permit safe passing. If in doubt she may sound the signals prescribed in Rule 34(*d*).

(ii) This Rule does not relieve the overtaking vessel of her obligation under Rule 13.

(*f*) A vessel nearing a bend or an area of a narrow channel or fairway where other vessels may be obscured by an intervening obstruction shall navigate with particular alertness and caution and shall sound the appropriate signal prescribed in Rule 34(*e*).

(*g*) Any vessel shall, if the circumstances of the case admit, avoid anchoring in a narrow channel.

RULE 10

Traffic separation schemes

(*a*) This Rule applies to traffic separation schemes adopted by the Organization:

(*b*) A vessel using a traffic separation scheme shall:

(i) proceed in the appropriate traffic lane in the general direction of traffic flow for that lane;

(ii) so far as practicable keep clear of a traffic separation line or separation zone;

(iii) normally join or leave a traffic lane at the termination of the lane, but when joining or leaving from either side shall do so at as small an angle to the general direction of traffic flow as practicable.

(*c*) A vessel shall so far as practicable avoid crossing traffic lanes, but if obliged to do so shall cross as nearly as practicable at right angles to the general direction of traffic flow.

(*d*) Inshore traffic zones shall not normally be used by through traffic which can safely use the appropriate traffic lane within the adjacent traffic separation scheme. However, vessels of less than 20 metres in length and sailing vessels may under all circumstances use inshore traffic zones.

(*e*) A vessel other than a crossing vessel or a vessel joining or leaving a lane shall not normally enter a separation zone or cross a separation line except:

(i) in cases of emergency to avoid immediate danger;

(ii) to engage in fishing within a separation zone.

(f) A vessel navigating in areas near the terminations of traffic separation schemes shall do so with particular caution.

(g) A vessel shall so far as practicable avoid anchoring in a traffic separation scheme or in areas near its terminations.

(h) A vessel not using a traffic separation scheme shall avoid it by as wide a margin as is practicable.

(i) A vessel engaged in fishing shall not impede the passage of any vessel following a traffic lane.

(j) A vessel of less than 20 metres in length or a sailing vessel shall not impede the safe passage of a power-driven vessel following a traffic lane.

(k) A vessel restricted in her ability to manoeuvre when engaged in an operation for the maintenance of safety of navigation in a traffic separation scheme is exempted from complying with this Rule to the extent necessary to carry out the operation.

(l) A vessel restricted in her ability to manoeuvre when engaged in an operation for the laying, servicing or picking up of a submarine cable, within a traffic separation scheme, is exempted from complying with this Rule to the extent necessary to carry out the operation.

Section II. Conduct of vessels in sight of one another

RULE 11

Application

Rules in this Section apply to vessels in sight of one another.

RULE 12

Sailing vessels

(a) When two sailing vessels are approaching one another, so as to involve risk of collision, one of them shall keep out of the way of the other as follows:

    (i) when each has the wind on a different side, the vessel which has the wind on the port side shall keep out of the way of the other;

    (ii) when both have the wind on the same side, the vessel which is to windward shall keep out of the way of the vessel which is to leeward;

    (iii) if a vessel with the wind on the port side sees a vessel to windward and cannot determine with certainty whether the other vessel has the wind on the port or on the starboard side, she shall keep out of the way of the other.

(b) For the purposes of this Rule the windward side shall be deemed to be the side opposite to that on which the mainsail is carried or, in the case of a square-rigged vessel, the side opposite to that on which the largest fore-and-aft sail is carried.

RULE 13

Overtaking

(a) Notwithstanding anything contained in the Rules of Part B, Sections I and II any vessel overtaking any other shall keep out of the way of the vessel being overtaken.

(*b*) A vessel shall be deemed to be overtaking when coming up with another vessel from a direction more than 22.5 degrees abaft her beam, that is, in such a position with reference to the vessel she is overtaking, that at night she would be able to see only the sternlight of that vessel but neither of her sidelights.

(*c*) When a vessel is in any doubt as to whether she is overtaking another, she shall assume that this is the case and act accordingly.

(*d*) Any subsequent alteration of the bearing between the two vessels shall not make the overtaking vessel a crossing vessel within the meaning of these Rules or relieve her of the duty of keeping clear of the overtaken vessel until she is finally past and clear.

RULE 14

Head-on situation

(*a*) When two power-driven vessels are meeting on reciprocal or nearly reciprocal courses so as to involve risk of collision each shall alter her course to starboard so that each shall pass on the port side of the other.

(*b*) Such a situation shall be deemed to exist when a vessel sees the other ahead or nearly ahead and by night she could see the mast head lights of the other in a line or nearly in a line and/or both sidelights and by day she observes the corresponding aspect of the other vessel.

(*c*) When a vessel is in any doubt as to whether such a situation exists she shall assume that it does exist and act accordingly.

RULE 15

Crossing situation

When two power-driven vessels are crossing so as to involve risk of collision, the vessel which has the other on her own starboard side shall keep out of the way and shall, if the circumstances of the case admit, avoid crossing ahead of the other vessel.

RULE 16

Action by give-way vessel

Every vessel which is directed to keep out of the way of another vessel shall, so far as possible, take early and substantial action to keep well clear.

RULE 17

Action by stand-on vessel

(*a*) (i) Where one of two vessels is to keep out of the way the other shall keep her course and speed.

(ii) The latter vessel may however take action to avoid collision by her manoeuvre alone, as soon as it becomes apparent to her that the vessel required to keep out of the way is not taking appropriate action in compliance with these Rules.

(b) When, from any cause, the vessel required to keep her course and speed finds herself so close that collision cannot be avoided by the action of the give-way vessel alone, she shall take such action as will best aid to avoid collision.

(c) A power-driven vessel which takes action in a crossing situation in accordance with sub-paragraph (a)(ii) of this Rule to avoid collision with another power-driven vessel shall, if the circumstances of the case admit, not alter course to port for a vessel on her own port side.

(d) This Rule does not relieve the give-way vessel of her obligation to keep out of the way.

RULE 18

Responsibilities between vessels

Except where Rules 9, 10 and 13 otherwise require:

(a) A power-driven vessel underway shall keep out of the way of:
  (i) a vessel not under command;
  (ii) a vessel restricted in her ability to manoeuvre;
  (iii) a vessel engaged in fishing;
  (iv) a sailing vessel.

(b) A sailing vessel underway shall keep out of the way of:
  (i) a vessel not under command;
  (ii) a vessel restricted in her ability to manoeuvre;
  (iii) a vessel engaged in fishing.

(c) A vessel engaged in fishing when underway shall, so far as possible, keep out of the way of:
  (i) a vessel not under command;
  (ii) a vessel restricted in her ability to manoeuvre.

(d) (i) Any vessel other than a vessel not under command or a vessel restricted in her ability to manoeuvre shall, if the circumstances of the case admit, avoid impeding the safe passage of a vessel constrained by her draught, exhibiting the signals in Rule 28;

  (ii) A vessel constrained by her draught shall navigate with particular caution having full regard to her special condition.

(e) A seaplane on the water shall, in general, keep well clear of all vessels and avoid impeding their navigation. In circumstances, however, where risk of collision exists, she shall comply with the Rules of this Part.

Section III. Conduct of vessels in restricted visibility

RULE 19

Conduct of vessels in restricted visibility

(a) This Rule applies to vessels not in sight of one another when navigating in or near an area of restricted visibility.

(*b*) Every vessel shall proceed at a safe speed adapted to the prevailing circumstances and conditions of restricted visibility. A power-driven vessel shall have her engines ready for immediate manoeuvre.

(*c*) Every vessel shall have due regard to the prevailing circumstances and conditions of restricted visibility when complying with the Rules of Section I of this Part.

(*d*) A vessel which detects by radar alone the presence of another vessel shall determine if a close-quarters situation is developing and/or risk of collision exists. If so, she shall take avoiding action in ample time, provided that when such action consists of an alteration of course, so far as possible the following shall be avoided:

> (i) an alteration of course to port for a vessel forward of the beam, other than for a vessel being overtaken;
>
> (ii) an alteration of course towards a vessel abeam or abaft the beam.

(*e*) Except where it has been determined that a risk of collision does not exist, every vessel which hears apparently forward of her beam the fog signal of another vessel, or which cannot avoid a close-quarters situation with another vessel forward of her beam, shall reduce her speed to the minimum at which she can be kept on her course. She shall if necessary take all her way off and in any event navigate with extreme caution until danger of collision is over.

PART C. LIGHTS AND SHAPES

RULE 20

Application

(*a*) Rules in this Part shall be complied with in all weathers.

(*b*) The Rules concerning lights shall be complied with from sunset to sunrise, and during such times no other lights shall be exhibited, except such lights as cannot be mistaken for the lights specified in these Rules or do not impair their visibility or distinctive character, or interfere with the keeping of a proper look-out.

(*c*) The lights prescribed by these Rules shall, if carried, also be exhibited from sunrise to sunset in restricted visibility and may be exhibited in all other circumstances when it is deemed necessary.

(*d*) The Rules concerning shapes shall be complied with by day.

(*e*) The lights and shapes specified in these Rules shall comply with the provisions of Annex I to these Regulations.

RULE 21

Definitions

(*a*) "Masthead light" means a white light placed over the fore and aft centreline of the vessel showing an unbroken light over an arc of the horizon of 225 degrees and so fixed as to show the light from right ahead to 22.5 degrees abaft the beam on either side of the vessel.

(*b*) "Sidelights" means a green light on the starboard side and a red light on the port side each showing an unbroken light over an arc of the horizon of 112.5 degrees and so fixed as to show the light from the right ahead to 22.5 degrees abaft the beam on its respective side. In a vessel of less than 20 metres in length the sidelights may be combined in one lantern carried on the fore and aft centreline of the vessel.

(c) "Sternlight" means a white light placed as nearly as practicable at the stern showing an unbroken light over an arc of the horizon of 135 degrees and so fixed as to show the light 67.5 degrees from right aft on each side of the vessel.

(d) "Towing light" means a yellow light having the same characteristics as the "sternlight" defined in paragraph (c) of this Rule.

(e) "All-round light" means a light showing an unbroken light over an arc of the horizon of 360 degrees.

(f) "Flashing light" means a light flashing at regular intervals at a frequency of 120 flashes or more per minute.

RULE 22

Visibility of lights

The lights prescribed in these Rules shall have an intensity as specified in Section 8 of Annex I to these Regulations so as to be visible at the following minimum ranges:

(a) In vessels of 50 metres or more in length:
  —a masthead light, 6 miles;

  —a sidelight, 3 miles;

  —a sternlight, 3 miles;

  —a towing light, 3 miles;

  —a white, red, green or yellow all-round light, 3 miles.

(b) In vessels of 12 metres or more in length but less than 50 metres in length:

  —a masthead light, 5 miles; except that where the length of the vessel is less than 20 metres, 3 miles;

  —a sidelight, 2 miles;

  —a sternlight, 2 miles;

  —a towing light, 2 miles;

  —a white, red, green or yellow all-round light, 2 miles.

(c) In vessels of less than 12 metres in length:

  —a masthead light, 2 miles;

  —a sidelight, 1 mile;

  —a sternlight, 2 miles;

  —a towing light, 2 miles;

  —a white, red, green or yellow all-round light, 2 miles.

(d) In inconspicuous, partly submerged vessels or objects being towed:

  —a white all-round light, 3 miles.

RULE 23

Power-driven vessels underway

(*a*) A power-driven vessel underway shall exhibit:

   (i) a masthead light forward;

   (ii) a second masthead light abaft of and higher than the forward one; except that a vessel of less than 50 metres in length shall not be obliged to exhibit such light but may do so;

   (iii) sidelights;

   (iv) a sternlight.

(*b*) An air-cushion vessel when operating in the non-displacement mode shall, in addition to the lights prescribed in paragraph (*a*) of this Rule, exhibit an all-round flashing yellow light.

(*c*) (i) A power-driven vessel of less than 12 metres in length may in lieu of the lights prescribed in paragraph (*a*) of this Rule exhibit an all-round white light and sidelights;

   (ii) a power-driven vessel of less than 7 metres in length whose maximum speed does not exceed 7 knots may in lieu of the lights prescribed in paragraph (*a*) of this Rule exhibit an all-round white light and shall, if practicable, also exhibit sidelights;

   (iii) the masthead light or all-round white light on a power-driven vessel of less than 12 metres in length may be displaced from the fore and aft centreline of the vessel if centreline fitting is not practicable, provided that the sidelights are combined in one lantern which shall be carried on the fore and aft centreline of the vessel or located as nearly as practicable in the same fore and aft line as the masthead light or the all-round white light.

RULE 24

Towing and pushing

(*a*) A power-driven vessel when towing shall exhibit:

   (i) instead of the light prescribed in Rule 23(*a*)(i) or (*a*)(ii), two masthead lights in a vertical line. When the length of the tow, measuring from the stern of the towing vessel to the after end of the tow exceeds 200 metres, three such lights in a vertical line;

   (ii) sidelights;

   (iii) a sternlight;

   (iv) a towing light in a vertical line above the sternlight;

   (v) when the length of the tow exceeds 200 metres, a diamond shape where it can best be seen.

(*b*) When a pushing vessel and a vessel being pushed ahead are rigidly connected in a composite unit they shall be regarded as a power-driven vessel and exhibit the lights prescribed in Rule 23.

(*c*) A power-driven vessel when pushing ahead or towing alongside, except in the case of a composite unit, shall exhibit:

   (i) instead of the light prescribed in Rule 23(*a*)(i) or (*a*)(ii), two masthead lights in a vertical line;

   (ii) sidelights;

   (iii) a sternlight.

(*d*) A power-driven vessel to which paragraph (*a*) or (*c*) of this Rule apply shall also comply with Rule 23(*a*)(ii).

(*e*) A vessel or object being towed, other than those mentioned in paragraph (*g*) of this Rule, shall exhibit:

   (i) sidelights;

   (ii) a sternlight;

   (iii) when the length of the tow exceeds 200 metres, a diamond shape where it can be best seen.

(*f*) Provided that any number of vessels being towed alongside or pushed in a group shall be lighted as one vessel:

   (i) a vessel being pushed ahead, not being part of a composite unit, shall exhibit at the forward end, sidelights;

   (ii) a vessel being towed alongside shall exhibit a sternlight and at the forward end, sidelights.

(*g*) An inconspicuous, partly submerged vessel or object, or combination of such vessels or objects being towed, shall exhibit:

   (i) if it is less than 25 metres in breadth, one all-round white light at or near the forward end and one at or near the after end except that dracones need not exhibit a light at or near the forward end;

   (ii) if it is 25 metres or more in breadth, two additional all-round white lights at or near the extremities of its breadth;

   (iii) if it exceeds 100 metres in length, additional all-round white lights between the lights prescribed in sub-paragraphs (i) and (ii) so that the distance between the lights shall not exceed 100 metres;

   (iv) a diamond shape at or near the aftermost extremity of the last vessel or object being towed and if the length of the tow exceeds 200 metres an additional diamond shape where it can best be seen and located as far forward as is practicable.

(*h*) Where from any sufficient cause it is impracticable for a vessel or object being towed to exhibit the lights or shapes prescribed in paragraph (*e*) or (*g*) of this Rule, all possible measures shall be taken to light the vessel or object towed or at least to indicate the presence of such vessel or object.

(*i*) where from any sufficient cause it is impracticable for a vessel not normally engaged in towing operations to display the lights prescribed in paragraph (*a*) or (*c*) of this Rule, such vessel shall not be required to exhibit those lights when engaged in towing another vessel in distress or otherwise in need of assistance. All possible measures shall be taken to indicate the nature of the relationship between the towing vessel and the vessel being towed as authorized by Rule 36, in particular by illuminating the towline.

RULE 25

Sailing vessels underway and vessels under oars

(*a*) A sailing vessel underway shall exhibit:

(i) sidelights;

(ii) a sternlight.

(*b*) In a sailing vessel of less than 20 metres in length the lights prescribed in paragraph (*a*) of this Rule may be combined in one lantern carried at or near the top of the mast where it can best be seen.

(*c*) A sailing vessel underway may, in addition to the lights prescribed in paragraph (*a*) of this Rule, exhibit at or near the top of the mast, where they can best be seen, two all-round lights in a vertical line, the upper being red and the lower green, but these lights shall not be exhibited in conjunction with the combined lantern permitted by paragraph (*b*) of this Rule.

(*d*) (i) A sailing vessel of less than 7 metres in length shall, if practicable, exhibit the lights prescribed in paragraph (*a*) or (*b*) of this Rule, but if she does not, she shall have ready at hand an electric torch or lighted lantern showing a white light which shall be exhibited in sufficient time to prevent collision.

(ii) A vessel under oars may exhibit the lights prescribed in this Rule for sailing vessels, but if she does not, she shall have ready at hand an electric torch or lighted lantern showing a white light which shall be exhibited in sufficient time to prevent collision.

(*e*) A vessel proceeding under sail when also being propelled by machinery shall exhibit forward where it can best be seen a conical shape, apex downwards.

RULE 26

Fishing vessels

(*a*) A vessel engaged in fishing, whether underway or at anchor, shall exhibit only the lights and shapes prescribed in this Rule.

(*b*) A vessel when engaged in trawling, by which is meant the dragging through the water of a dredge net or other apparatus used as a fishing appliance, shall exhibit:

(i) two all-round lights in a vertical line, the upper being green and the lower white, or a shape consisting of two cones with their apexes together in a vertical line one above the other; a vessel of less than 20 metres in length may instead of this shape exhibit a basket;

(ii) a masthead light abaft of and higher than the all-round green light; a vessel of less than 50 metres in length shall not be obliged to exhibit such a light but may do so;

(iii) when making way through the water, in addition to the lights prescribed in this paragraph, sidelights and a sternlight.

166

(*c*) A vessel engaged in fishing, other than trawling, shall exhibit:

  (i) two all-round lights in a vertical line, the upper being red and the lower white, or a shape consisting of two cones with apexes together in a vertical line one above the other; a vessel of less than 20 metres in length may instead of this shape exhibit a basket;

  (ii) when there is outlying gear extending more than 150 metres horizontally from the vessel, an all-round white light or a cone apex upwards in the direction of the gear;

  (iii) when making way through the water, in addition to the lights prescribed in this paragraph, sidelights and a sternlight.

(*d*) A vessel engaged in fishing in close proximity to other vessels engaged in fishing may exhibit the additional signals described in Annex II to these Regulations.

(*e*) A vessel when not engaged in fishing shall not exhibit the lights or shapes prescribed in this Rule, but only those prescribed for a vessel of her length.

RULE 27

Vessels not under command or restricted in their ability to manoeuvre

(*a*) A vessel not under command shall exhibit:

  (i) two all-round red lights in a vertical line where they can best be seen;

  (ii) two balls or similar shapes in a vertical line where they can best be seen;

  (iii) when making way through the water, in addition to the lights prescribed in this paragraph, sidelights and a sternlight.

(*b*) A vessel restricted in her ability to manoeuvre, except a vessel engaged in mineclearance operations, shall exhibit:

  (i) three all-round lights in a vertical line where they can best be seen. The highest and lowest of these lights shall be red and the middle light shall be white;

  (ii) three shapes in a vertical line where they can best be seen. The highest and lowest of these shapes shall be balls and the middle one a diamond;

  (iii) when making way through the water, a masthead light or lights, sidelights and a sternlight, in addition to the lights prescribed in sub-paragraph (i);

  (iv) when at anchor, in addition to the lights or shapes prescribed in sub-paragraphs (i) and (ii), the light, lights or shape prescribed in Rule 30.

(*c*) A power-driven vessel engaged in a towing operation such as severely restricts the towing vessel and her tow in their ability to deviate from their course shall, in addition to the lights or shapes prescribed in Rule 24(*a*), exhibit the lights or shapes prescribed in sub-paragraphs (*b*)(i) and (ii) of this Rule.

(*d*) A vessel engaged in dredging or underwater operations, when restricted in her ability to manoeuvre, shall exhibit the lights and shapes prescribed in sub-paragraphs (*b*)(i), (ii) and (iii) of this Rule and shall in addition, when an obstruction exists, exhibit:

  (i) two all-round red lights or two balls in a vertical line to indicate the side on which the obstruction exists;

  (ii) two all-round green lights or two diamonds in a vertical line to indicate the side on which another vessel may pass;

(iii) when at anchor, the lights or shapes prescribed in this paragraph instead of the lights or shape prescribed in Rule 30.

(*e*) Whenever the size of a vessel engaged in diving operations makes it impracticable to exhibit all lights and shapes prescribed in paragraph (*d*) of this Rule, the following shall be exhibited:

(i) three all-round lights in a vertical line where they can best be seen. The highest and lowest of these lights shall be red and the middle light shall be white;

(ii) a rigid replica of the International Code flag "A" not less than 1 metre in height. Measures shall be taken to ensure its all-round visibility.

(*f*) A vessel engaged in mineclearance operations shall in addition to the lights prescribed for a power-driven vessel in Rule 23 or to the lights or shape prescribed for a vessel at anchor in Rule 30 as appropriate, exhibit three all-round green lights or three balls. One of these lights or shapes shall be exhibited near the foremast head and one at each end of the fore yard. These lights or shapes indicate that it is dangerous for another vessel to approach within 1000 metres of the mineclearance vessel.

(*g*) Vessels of less than 12 metres in length, except those engaged in diving operations, shall not be required to exhibit the lights and shapes prescribed in this Rule.

(*h*) The signals prescribed in this Rule are not signals of vessels in distress and requiring assistance. Such signals are contained in Annex IV to these Regulations.

RULE 28

Vessels constrained by their draught

A vessel constrained by her draught may, in addition to the lights prescribed for power-driven vessels in Rule 23, exhibit where they can best be seen three all-round red lights in a vertical line, or a cylinder.

RULE 29

Pilot vessels

(*a*) A vessel engaged on pilotage duty shall exhibit:

(i) at or near the masthead, two all-round lights in a vertical line, the upper being white and the lower red;

(ii) when underway, in addition, sidelights and a sternlight;

(iii) when at anchor, in addition to the lights prescribed in sub-paragraph (i), the light, lights or shape prescribed in Rule 30 for vessels at anchor.

(*b*) A pilot vessel when not engaged on pilotage duty shall exhibit the lights or shapes prescribed for a similar vessel of her length.

RULE 30

Anchored vessels and vessels aground

(*a*) A vessel at anchor shall exhibit where it can best be seen
    (i) in the fore part, an all-round white light or one ball;
    (ii) at or near the stern and at a lower level than the light prescribed in sub-paragraph (i), an all-round white light.

(*b*) A vessel of less than 50 metres in length may exhibit an all-round white light where it can best be seen instead of the lights prescribed in paragraph (*a*) of this Rule.

(*c*) A vessel at anchor may, and a vessel of 100 metres and more in length shall, also use the available working or equivalent lights to illuminate her decks.

(*d*) A vessel aground shall exhibit the lights prescribed in paragraph (*a*) or (*b*) of this Rule and in addition, where they can best be seen:
    (i) two all-round red lights in a vertical line;
    (ii) three balls in a vertical line.

(*e*) A vessel of less than 7 metres in length, when at anchor, not in or near a narrow channel, fairway or anchorage, or where other vessels normally navigate, shall not be required to exhibit the lights or shape prescribed in paragraphs (*a*)and (*b*) of this Rule.

(*f*) A vessel of less than 12 metres in length, when aground, shall not be required to exhibit the lights or shapes prescribed in sub-paragraphs (*d*)(i) and (ii) of this Rule.

RULE 31

Seaplanes

Where it is impracticable for a seaplane to exhibit lights and shapes of the characteristics or in the positions prescribed in the Rules of this Part she shall exhibit lights and shapes as closely similar in characteristics and position as is possible.

PART D. SOUND and LIGHT SIGNALS

RULE 32

Definitions

(*a*) The word "whistle" means any sound signalling appliance capable of producing the prescribed blasts and which complies with the specifications in Annex III to these Regulations.

(*b*) The term "short blast" means a blast of about one second's duration.

(*c*) The term "prolonged blast" means a blast of from four to six seconds' duration.

RULE 33

Equipment for sound signals

(*a*) A vessel of 12 metres or more in length shall be provided with a whistle and a bell and a vessel of 100 metres or more in length shall, in addition, be provided with a gong, the tone and sound of which cannot be confused with that of the bell. The whistle, bell and gong shall comply with the specifications in Annex III to these Regulations. The bell or gong or both may be replaced by other equipment having the same respective sound characteristics, provided that manual sounding of the prescribed signals shall always be possible.

(b) A vessel of less than 12 metres in length shall not be obliged to carry the sound signalling appliances prescribed in paragraph (a) of this Rule but if she does not, she shall be provided with some other means of making an efficient sound signal.

RULE 34

Manoeuvring and warning signals

(a) When vessels are in sight of one another, a power-driven vessel underway, when manoeuvring as authorized or required by these Rules, shall indicate that manoeuvre by the following signals on her whistle:

—one short blast to mean "I am altering my course to starboard";

—two short blasts to mean "I am altering my course to port";

—three short blasts to mean "I am operating astern propulsion".

(b) Any vessel may supplement the whistle signals prescribed in paragraph (a) of this Rule by light signals, repeated as appropriate, whilst the manoeuvre is being carried out:

(i) these light signals shall have the following significance:

—one flash to mean "I am altering my course to starboard";

—two flashes to mean "I am altering my course to port";

—three flashes to mean "I am operating astern propulsion";

(ii) the duration of each flash shall be about one second, the interval between flashes shall be about one second, and the interval between successive signals shall be not less than ten seconds;

(iii) the light used for this signal shall, if fitted, be an all-round white light, visible at a minimum range of 5 miles, and shall comply with the provisions of Annex I to these Regulations.

(c) When in sight of one another in a narrow channel or fairway:

(i) a vessel intending to overtake another shall in compliance with Rule 9 (e)(i) indicate her intention by the following signals on her whistle:

—two prolonged blasts followed by one short blast to mean "I intend to overtake you on your starboard side";

—two prolonged blasts followed by two short blasts to mean "I intend to overtake you on your port side";

(ii) the vessel about to be overtaken when acting in accordance with Rule 9(e)(i) shall indicate her agreement by the following signal on her whistle:

—one prolonged, one short, one prolonged and one short blast, in that order.

(d) When vessels in sight of one another are approaching each other and from any cause either vessel fails to understand the intentions or actions of the other, or is in doubt whether sufficient action is being taken by the other to avoid collision, the vessel in doubt shall immediately indicate such doubt by giving at least five short and rapid blasts on the whistle. Such signal may be supplemented by a light signal of at least five short and rapid flashes.

(e) A vessel nearing a bend or an area of a channel or fairway where other vessels may be obscured by an intervening obstruction shall sound one prolonged blast. Such signal shall be answered with a prolonged blast by any approaching vessel that may be within hearing around the bend or behind the intervening obstruction.

(f) If whistles are fitted on a vessel at a distance apart of more than 100 metres, one whistle only shall be used for giving manoeuvring and warning signals.

RULE 35

Sound signals in restricted visibility

In or near an area of restricted visibility, whether by day or night, the signals prescribed in this rule shall be used as follows:

(a) A power-driven vessel making way through the water shall sound at intervals of not more than 2 minutes one prolonged blast.

(b) A power-driven vessel underway but stopped and making no way through the water shall sound at intervals of not more than 2 minutes two prolonged blasts in succession with an interval of about 2 seconds between them.

(c) A vessel not under command, a vessel restricted in her ability to manoeuvre, a vessel constrained by her draught, a sailing vessel, a vessel engaged in fishing and a vessel engaged in towing or pushing another vessel shall, instead of the signals prescribed in paragraphs (a) or (b) of this Rule, sound at intervals of not more than 2 minutes three blasts in succession, namely one prolonged followed by two short blasts.

(d) A vessel engaged in fishing, when at anchor, and a vessel restricted in her ability to manoeuvre when carrying out her work at anchor, shall instead of the signals prescribed in paragraph (g) of this Rule sound the signal prescribed in paragraph (c) of this Rule.

(e) A vessel towed or if more than one vessel is towed the last vessel of the tow, if manned, shall at intervals of not more than 2 minutes sound four blasts in succession, namely one prolonged followed by three short blasts. When practicable, this signal shall be made immediately after the signal made by the towing vessel.

(f) When a pushing vessel and a vessel being pushed ahead are rigidly connected in a composite unit they shall be regarded as a power-driven vessel and shall give the signals prescribed in paragraphs (a) or (b) of this Rule.

(g) A vessel at anchor shall at intervals of not more than one minute ring the bell rapidly for about 5 seconds. In a vessel of 100 metres or more in length the bell shall be sounded in the forepart of the vessel and immediately after the ringing of the bell the gong shall be sounded rapidly for about 5 seconds in the after part of the vessel. A vessel at anchor may in addition sound three blasts in succession, namely one short, one prolonged and one short blast, to give warning of her position and of the possibility of collision to an approaching vessel.

(h) A vessel aground shall give the bell signal and if required the gong signal prescribed in paragraph (*g*) of this Rule and shall, in addition, give three separate and distinct strokes on the bell immediately before and after the rapid ringing of the bell. A vessel aground may in addition sound an appropriate whistle signal.

(*i*) A vessel of less than 12 metres in length shall not be obliged to give the above-mentioned signals but, if she does not, shall make some other efficient sound signal at intervals of not more than 2 minutes.

(*j*) A pilot vessel when engaged on pilotage duty may in addition to the signals prescribed in paragraphs (*a*), (*b*) or (*g*) of this Rule sound an identity signal consisting of four short blasts.

RULE 36

Signals to attract attention

If necessary to attract the attention of another vessel any vessel may make light or sound signals that cannot be mistaken for any signal authorized elsewhere in these Rules, or may direct the beam of her searchlight in the direction of the danger, in such a way as not to embarrass any vessel. Any light to attract the attention of another vessel shall be such that it cannot be mistaken for any aid to navigation. For the purpose of this Rule the use of high intensity intermittent or revolving lights, such as strobe lights, shall be avoided.

RULE 37

Distress signals

When a vessel is in distress and requires assistance she shall use or exhibit the signals described in Annex IV to these Regulations.

PART E. EXEMPTIONS

RULE 38

Exemptions

Any vessel (or class of vessels) provided that she complies with the requirements of the International Regulations for Preventing Collisions at Sea, 1960(*a*), the keel of which is laid or which is at a corresponding stage of construction before the entry into force of these Regulations may be exempted from compliance therewith as follows:

(*a*) The installation of lights with ranges prescribed in Rule 22, until four years after the date of entry into force of these Regulations.

(*b*) The installation of lights with colour specifications as prescribed in Section 7 of Annex I to these Regulations, until four years after the date of entry into force of these Regulations.

(*c*) The repositioning of lights as a result of conversion from Imperial to metric units and rounding off measurement figures, permanent exemption.

(*d*) (i) The repositioning of masthead lights on vessels of less than 150 metres in length, resulting from the prescriptions of Section 3(*a*) of Annex I to these Regulations, permanent exemption.

    (ii) The repositioning of masthead lights on vessels of 150 metres or more in length, resulting from the prescriptions of Section 3(*a*) of Annex I to these Regulations, until nine years after the date of entry into force of these Regulations.

(*e*) The repositioning of masthead lights resulting from the prescriptions of Section 2(*b*) of Annex I to these Regulations, until nine years after the date of entry into force of these Regulations.

(*f*) The repositioning of sidelights resulting from the prescriptions of Sections 2(*g*) and 3(*b*) of Annex I to these Regulations, until nine years after the date of entry into force of these Regulations.

(*g*) The requirements for sound signal appliances prescribed in Annex III to these Regulations, until nine years after the date of entry into force of these Regulations.

(*h*) The repositioning of all-round lights resulting from the prescription of Section 9(*b*) of Annex I to these Regulations, permanent exemption.

ANNEX I

Positioning and technical details of lights and shapes

In this Annex are to be found all the technical specifications governing lights and shapes.

It concludes by stating that the construction of lights and shapes on board the vessel shall be to the satisfaction of the appropriate authority of the State whose flag the vessel is entitled to fly.

Of direct practical concern to fishermen, as opposed to boatbuilders, is the following Rule:

6. *Shapes*

(*a*) Shapes shall be black and of the following sizes:

    (i) a ball shall have a diameter of not less than 0·6 metre;

    (ii) a cone shall have a base diameter of not less than 0·6 metre and a height equal to its diameter;

    (iii) a cylinder shall have a diameter of at least 0·6 metre and a height of twice its diameter;

    (iv) a diamond shape shall consist of two cones as defined in (ii) above having a common base.

(*b*) The vertical distance between shapes shall be at least 1·5 metres.

(*c*) In a vessel of less than 20 metres in length shapes of lesser dimensions but commensurate with the size of the vessel may be used and the distance apart may be correspondingly reduced.

ANNEX II

Additional signals for fishing vessels fishing in close proximity

1. *General*

The lights mentioned herein shall, if exhibited in pursuance of Rule 26(*d*), be placed where they can best be seen. They shall be at least 0·9 metre apart but at a lower level than lights prescribed in Rule 26(*b*)(i) and (*c*)(i). The lights shall be visible all round the horizon at a distance of at least 1 mile but at a lesser distance than the lights prescribed by these Rules for fishing vessels.

2. *Signals for trawlers*

(*a*) Vessels when engaged in trawling, whether using demersal or pelagic gear, may exhibit:

  (i) when shooting their nets:
       two white lights in a vertical line;

  (ii) when hauling their nets:
       one white light over one red light in a vertical line;

  (iii) when the net has come fast upon an obstruction:
       two red lights in a vertical line.

(*b*) Each vessel engaged in pair trawling may exhibit:

  (i) by night, a searchlight directed forward and in the direction of the other vessel of the pair;

  (ii) when shooting or hauling their nets or when their nets have come fast upon an obstruction, the lights prescribed in 2(*a*) above.

3. *Signals for purse seiners*

Vessels engaged in fishing with purse seine gear may exhibit two yellow lights in a vertical line. These lights shall flash alternately every second and with equal light and occultation duration. These lights may be exhibited only when the vessel is hampered by its fishing gear.

ANNEX III

Technical details of sound signal appliances

In this Annex are to be found all the details concerning whistles, bells and gongs. It concludes by stating that the construction of sound signal appliances, their performance and their installation on board the vessel shall be to the satisfaction of the appropriate authority of the State whose flag the vessel is entitled to fly.

ANNEX IV

Distress signals

1. The following signals, used or exhibited either together or separately, indicate distress and need of assistance:

174

(a) a gun or other explosive signal fired at intervals of about a minute;

(b) a continuous sounding with any fog-signalling apparatus;

(c) rockets or shells, throwing red stars fired one at a time at short intervals;

(d) a signal made by radiotelegraphy or by any other signalling method consisting of the group $\cdots - - - \cdots$ (SOS) in the Morse Code;

(e) a signal sent by radiotelephony consisting of the spoken word "Mayday";

(f) the International Code Signal of distress indicated by N.C.;

(g) a signal consisting of a square flag having above or below it a ball or anything resembling a ball;

(h) flames on the vessel (as from a burning tar barrel, oil barrel, etc.);

(i) a rocket parachute flare or a hand flare showing a red light;

(j) a smoke signal giving off orange-coloured smoke;

(k) slowly and repeatedly raising and lowering arms outstretched to each side;

(l) the radiotelegraph alarm signal;

(m) the radiotelephone alarm signal;

(n) signals transmitted by emergency position-indicating radio beacons.

2. The use or exhibition of any of the foregoing signals except for the purpose of indicating distress and need of assistance and the use of other signals which may be confused with any of the above signals is prohibited.

3. Attention is drawn to the relevant sections of the International Code of Signals, the Merchant Ship Search and Rescue Manual and the following signals:

(a) a piece of orange-coloured canvas with either a black square and circle or other appropriate symbol (for identification from the air);

(b) a dye marker.

16 International code of signals

Single letter signals by flag (see *page 312*)

A. I have a diver down. Keep well clear at slow speed.
*B. I am taking in, or discharging, or carrying explosives.
*C. Yes, 'affirmative', or the 'significance of the previous group should be read in the affirmative'.
*D. Keep clear of me. I am manoeuvring with difficulty.
*E. I am altering my course to starboard.
F. I am disabled. Communicate with me.
*G. I require a pilot.
*H. I have a pilot on board.
*I. I am altering my course to port.
J. I am on fire and have dangerous cargo on board.
K. I wish to communicate with you.
L. You should stop your vessel instantly.
*M. My vessel is stopped and making no way through the water.
N. No, negative or 'the significance of the previous group should be read in the negative'.
O. Man overboard.
P. In harbour, hoisted at the foremast. All persons should report on board 'I am about to sail'.
Q. My vessel is healthy and I require pratique.
R. Not allocated.
*S. My engines are going astern.
*T. Keep clear of me, I am engaged in pair trawling.
U. You are running into danger.
V. I require assistance.
W. I require medical assistance.
X. Stop carrying out your intentions and watch for my signals.
Y. I am dragging my anchor.
*Z. I require a tug. (When made by fishing vessels operating in close proximity, it means 'I am shooting my nets'.)
These signals may be made by any method of signalling. Those marked by

an asterisk may only be made in compliance with the *International Regulations for Preventing Collisions at Sea*, Rules 34 and 35.

Morse Code

Alphabet   Numerals

Alphabet		Alphabet		Numerals	
A	· –	N	– ·	1	· – – – –
B	– · · ·	O	– – –	2	· · – – –
C	– · – ·	P	· – – ·	3	· · · – –
D	– · ·	Q	– – · –	4	· · · · –
E	·	R	· – ·	5	· · · · ·
F	· · – ·	S	· · ·	6	– · · · ·
G	– – ·	T	–	7	– – · · ·
H	· · · ·	U	· · –	8	– – – · ·
I	· ·	V	· · · –	9	– – – – ·
J	· – – –	W	· – –	0	– – – – –
K	– · –	X	– · · –		
L	· – · ·	Y	– · – –		
M	– –	Z	– – · ·		

The 1972 *International Regulations for Preventing Collisions* at Sea provide for the following signals to be made by sound only. Those signals with an asterisk may be supplemented by a light signal as prescribed by Rules 34 and 35.

Rule 34

*E	·	One short blast, I am altering my course to starboard.
*I	· ·	Two short blasts, I am altering my course to port.
*S	· · ·	Three short blasts, I am operating astern propulsion.
G	– – ·	I intend to overtake you on your starboard side (In a channel or fairway)
Z	– – · ·	I intend to overtake you on your port side (In a channel or fairway)
C	– · – ·	Indicates agreement to be overtaken and in reply to G or Z (In channel or fairway)
*5	· · · · ·	To be sounded when vessels are approaching one another, under circumstances whereby doubt exists or there is failure to understand the intentions or actions of the other and risk of collision develops.
T	–	One prolonged blast may be sounded by a vessel about to round a bend in a fairway or channel which is obscured by the intervening land. A vessel approaching the bend from the opposite direction, on hearing the prolonged blast may answer with the same signal.

Rule 35. Restricted visibility

T –   Prolonged blast, sounded at intervals of not more than two minutes by a power driven vessel making way.

M – –   Two prolonged blasts sounded at intervals of not more than two minutes by a power driven vessel under way, but stopped and not making way through the water.

D – · ·   Sounded by a vessel not under command, a vessel restricted in her ability to manoeuvre, a vessel constrained by her draught, a sailing vessel, a vessel engaged in fishing, a vessel engaged in towing or pushing.

B – · · ·   A vessel towed, or if more than one vessel is towed, the last vessel towed (if manned) shall sound this signal. This signal should be sounded immediately after the signal made by the towing vessel.

R · – ·   This signal may be made by a vessel at anchor, in addition to the ringing of the bell, but only on the approach of another vessel in order to warn her of the anchored vessel's position and the risk of collision.

H · · · ·   A pilot vessel engaged on pilotage duty, whether under way or at anchor, may in poor visibility sound this signal in addition to the normal fog signals.

Rule 35 requires that a vessel aground shall give the normal rapid ringing of the bell for about five seconds at intervals of not more than one minute, and then give three separate and distinct strokes on the bell. It then says that the grounded vessel may in addition sound an appropriate whistle signal.

The advice on sounding an appropriate whistle signal might appear to be ambiguous. If the vessel is aground in fog, there is no appropriate whistle signal, so that Rule 36[*] must be considered. The warning signal of five short blasts may be given, Rule 34(d), but the vessels would not be in sight of one another as required by the rule.

In most ports there are local bye-laws in which there is given a whistle signal to be sounded by a vessel aground, on the approach of another vessel in poor visibility. If in doubt an appropriate signal to sound might well be the letter U, which is the single letter flag signal, meaning 'You are running into danger'. This signal has been used by light and sound for many years especially when a vessel has been seen to be running into danger.

---

[*]RULE 36 *Signals to attract attention*

If necessary to attract the attention of another vessel any vessel may make light or sound signals that cannot be mistaken for any signal authorized elsewhere in these Rules, or may direct the beam of her searchlight in the direction of the danger, in such a way as not to embarrass any vessel. Any light to attract attention of another vessel shall be such that it cannot be mistaken for any aid to navigation. For the purpose of this Rule the use of high intensity intermittent or revolving lights, such as strobe lights, shall be avoided.

17 Submarine telegraph cables and offshore installations – safety zones

There are many telephone and telegraph cables which lie on the seabed. These submarine cables are subject to damage by ships' anchors, but bottom trawls are mainly responsible for a considerable amount of damage to these cables and the cost of repairs is very high.

There are three important reasons why trawlers should keep clear of cables areas when towing gear:

— Damage to modern telephone cables can cause serious disruption to world communications affecting safety of life at sea, in the air, trade, international business and affairs. Each cable is capable of carrying thousands of telephone and other messages at the same time.
— Most modern submarine cables carry high voltages which can prove lethal if attempts are made to cut or chop them.
— Loss of gear, time and in some cases a valuable catch may result if a trawler fouls a submarine cable.

Law of the sea relating to submarine cables

The International Convention for the Protection of Submarine Cables 1884, as extended by the Convention on the High Seas 1958, and agreed by member nations, stipulates that:

(a) Vessels shall not remain or close within one mile of vessels engaged in laying or repairing submarine cables or pipelines and vessels engaged in such work shall exhibit the signals laid down in the *International Regulations for Preventing Collisions at Sea 1972.* (Rule 27).
Fishing gear and nets shall also be removed to or kept at a distance of one mile from vessels showing these signals, but fishing vessels shall be allowed 24 hours after the first signal is visible to them to get clear.
(b) Buoys marking cables and pipelines shall not be approached within 1/4 mile and fishing gear and nets shall be kept at the same distance from them.
(c) It is an offence for any person to deliberately, or through culpable negligence, damage or break a submarine cable and such a person is liable to a penalty of imprisonment or a fine or both. It is also obligatory

Fig 17.1 Approximate position of submarine cables normally shown as magenta coloured wavy lines on Admiralty charts

that anyone who fouls a cable must sacrifice his gear rather than cut the cable.

(d) Owners and skippers of fishing vessels who can prove that they have sacrificed an anchor or fishing gear, in order to avoid damaging a submarine cable or pipeline, shall receive compensation. A claim for such compensation should be made within 24 hours of arrival in port to the appropriate authority giving full particulars. It is also advisable to report the incident by radio. Give the following information:

— The date and time of incident.
— The exact position, shore bearings or readings by electronic navigation system.
— Depth of water and description of cable if sighted.

In the UK, the appropriate authorities are the DOT, Customs and Excise, Coastguard, and Fishery Officer. An entry made in the deck log and a statement supported by evidence of the crew should be drawn up immediately after the occurrence so as to support the claim.

Charts and position of cables

On the Admiralty navigation charts used by British fishing vessels the international symbol indicating a submarine cable is shown as a corrugated or wavy line coloured magenta. On other charts the cable may be shown as a corrugated black line. Particular importance is attached to showing cables on fishing charts obtained from other sources.

In coastal areas normally fished by trawlers charting accuracy at about 100 miles offshore is usually better than 2½ cables, improving to about one cable close inshore. Cables laid in mid-ocean prior to 1970 may be as much as two miles from their charted position, but present day mid-ocean accuracy is usually better than ½ mile.

On very old charts which are still in existence, cables may not be shown. It has already been pointed out elsewhere in this book that skippers and mates should always use the most up-to-date and largest scale chart for navigation. We now have another reason, the submarine cable, which will be more clearly and accurately defined on the large scale coastal chart, than on the small.

In the UK, if in any doubt as to the location of cables in a particular fishing area, application may be made to the undermentioned address for cable warning charts covering the area of interest.

Telecommunications HQ
Marine Division,
Central Marine Depot,
Berth 203, Western Docks,
Southampton SO1 0HH.

Prevention of damage to submarine cables

Practically all cable breaks resulting from fishing are caused by otter boards or beam trawls. In most cases the cable is broken under tension without being brought to the surface. In some cases the cable armouring wires become severed and rucked up as the door or beam trawl is towed across and the broken wires then become entangled with the net. In a few cases the cable is broken but one end remains foul of the gear and is brought on board. With the more powerful trawler the cable may be brought on board and cases are known where the cable has been chopped or burnt through in order to clear.

The most obvious method by which to avoid cable fouling is to not knowingly fish over a submarine cable.

However if the trawl door is well maintained and constructed and providing that the cable conforms closely to the contours of the seabed, fouling will not normally occur. But any of the following circumstances or a combination of these, may result in fouling.

— Poorly maintained trawl doors and fittings
— Cable suspended over bottom depressions
— Cable turns or bight standing proud of the bottom due to excessive slack resulting from repairs in the vicinity
— Door lying flat on the bottom when the trawler has turned too quickly
— Trawl net snagging on broken cable armour wires, as a result of trawl doors fouling the cable.

Trawl doors used in demersal fishing are designed with a rounded leading edge so that a submarine cable which lies on the seabed will, in general, allow the door to pass over it without fastening. However, when a cable is not resting on the bottom at all points, the probability of fouling with a door may be considerable.

Skippers and mates will appreciate that a trawl door is designed to maintain the spread at the mouth of the trawl. The brackets are so constructed that a sheering angle of about $30°$ outwards from the ship's course is maintained. There is of course a second component of force and that is the vertical moment which positions the door above the seabed. The position of the net and door is determined by the speed at which the ship tows and the length of warp paid out, ie the distance between the ship's towing block and the codend. The length of warp paid out and the speed of the ship when demersal fishing is usually adjusted so that the trawl doors are either on the bottom or very close to the bottom. The vertical angle of the board relative to the seabed depends on the position of the towing bracket. Usually the bracket is midway between the top and bottom of the door and the door is towed close to or along the bottom depending on the speed or catenary of the warp.

It will be seen from this data that the leading edge of a trawl door will very likely catch on a cable if the cable is in suspension over a depression and the tow is being made at approximately right angles to the line of the cable. The door will be forced over on to its side by reason of the pull on the towing bracket and the submarine cable will very likely jam between door and warp at the towing bracket.

In the event that the door passes over the cable, it is nearly certain that the bridles, footropes and ground weights or steel bobbins will foul or strike the cable causing damage to the fishing gear and the cable.

The beam trawl, by its construction and mode of operation, will almost be certain to foul a cable which is in suspension. The head of the beam trawl, with its various attachments, being dragged along the seabed will always be a danger to cables which are proud of the bottom.

The following advice on avoidance of cable damage should be followed:

— Do not knowingly trawl in the vicinity of cables.
— Do not alter course too quickly when trawling so that the trawl door on the inside leg turns over and results in it being dragged slowly along the sea bed.
— See that all bolt heads on the inner side of the door are round and smooth.
— Nuts, if used, should be on the outer side of the board and project as little as possible, also have their corners rounded off.
— The bottom of the board should not be straight throughout its length but should be well curved towards the fore end.
— The shoe should be smooth and the bolts which secure it to the board should be counter-sunk.
— The fore end of the shoe should be carried well up the leading edge of the board, should fit closely to the iron plates running up the front edge, and should terminate without either projection or indentation.
— Generally, all attachments should be as simple and smooth as possible and on every part of the door the aim should be to afford no resting place for a cable if it should be accidentally picked up.
— The attention of skippers and mates is drawn to the importance of placing all trawl gear shackles so that the bow of the shackle faces the direction of travel.

Action to be taken if a cable is fouled

When a cable is fouled, great care must be taken in attempting to free the trawl gear. Comparatively little pressure is required to force the armouring wires through the insulation on to the copper conductor. As soon as this happens water will reach the centre conductor and the cable will be rendered unusable even though it may not be actually broken.

If the gear cannot be freed without risk of damage to the cable, then it should be abandoned and the loss claimed for as previously described.

In the event that a submarine cable is brought up either on the gear or on the flukes of an anchor and it is possible to free the cable, then a slip rope should be passed under the cable and brought back on board. A wire should not be used for this purpose. The anchor can then be lowered clear of the cable or the gear can be either lowered clear or brought on board, and the cable slipped.

The ship should be handled so that the cable does not slide along the flukes of the anchor or if caught on the gear, so that sliding or chafing does not take place. Try to avoid damage to the cable armouring and insulation.

Under no circumstances should the cable be chopped or burnt off. High voltages are fed into certain submarine cables other than power transmission cables. Serious risk to life exists due to electric shock or severe injury due to burns if any attempt to cut the cable is made. No claims in respect of injury, loss or damage suffered through interference with the cable will be accepted.

Offshore installations – safety zones

Under international law a coastal state may construct and maintain on the continental shelf, installations and other devices necessary for the exploration and exploitation of its natural resources, establish safety zones around such installations, and take within these zones measures for their protection. Safety zones may extend to a distance of 500 metres around installations, measured from their outer edges. *Ships of all nationalities are required to respect these safety zones.*

In their national laws many coastal states have made entry by unauthorised vessels into declared zones a criminal offence. As the type of installation subject to safety zones varies from state to state, *mariners are advised always to assume the existence of a safety zone* unless they have information to the contrary. Installations around which safety zones may have been established include fixed platforms, mobile rigs (while on station), tanker loading moorings, and such sea-bed installations as submerged wellheads.

Installations on the British continental shelf (and in British tidal and territorial waters) are protected by safety zones promulgated by Orders under the Oil and Gas (Enterprise) Act 1982. The existence of safety zones established by these Orders is promulgated in Notices to Mariners or, in the case of mobile drilling rigs, by radio navigational warnings in the NAV AREA ONE series. Entry into a safety zone by an unauthorised vessel makes the owner, master, and others who may have contributed to the offence, liable to a fine or imprisonment, or both. The Orders may apply to parts of installations, where the latter are being assembled or dismantled; they do not apply to submarine pipe-lines.

Areas extending beyond 500 metres from installations

Certain fields which are being developed, or are currently producing oil or gas, are designated **Development Areas.** Within these areas, the limits of which are charted, there are likely to be construction and maintenance vessels including submarine craft, divers and obstructions, possibly marked by buoys. Supply vessels and, in some cases, tankers, frequently manoeuvre within these fields. *Mariners are strongly advised to keep outside such areas.* Some coastal states have declared prohibitions on entry into, or on fishing and anchoring within, areas extending beyond 500 metres from installations. Publication of the details of such wider areas is solely for the safety and convenience of shipping, and implies no recognition of the international validity of such restrictions.

Part IV – Fishing operations

18 Trawling

It is not the intention here to describe the various methods of fishing and the techniques employed, which are well covered by a variety of publications, in particular 'Commercial Fishing Methods' (*see useful publications on page 320*).

Rather it is considered useful in a handbook of this kind to elaborate on certain aspects experienced in such active methods as bottom trawling, including beam trawling, seine netting and scallop dredging, and also to point out what other fishing vessels must be aware of in the presence of those operating static gear.

Demersal and pelagic trawling

Demersal is the name given to those varieties of fish which live and feed close to the bottom of the sea. They may be caught by using a ground, bottom or demersal trawl and the species include cod, haddock, flatfish, berghylt, coalfish, *etc*. (See *Fig 18.1*).

Pelagic is the name given to those varieties of fish which live and feed in mid-water, *ie* herring, mackerel, sprats, *etc*. They are not usually readily caught by a ground trawl and therefore a pelagic or mid-water trawl has been developed. (See *Fig 18.2*).

'Delagic' trawls are basically mid-water trawls, the design of which incorporates features that allow the net to be fished demersally for those fish that are close to the bottom or pelagically for herring, mackerel, sprats, *etc*, without the need to change from one fishing mode to the other. They are more robust than pelagic trawls.

The assessment of a fishing situation may require a skipper to change over from one mode of fishing to another, that is, from bottom to mid-water or vice-versa. This would normally present the skipper and crew with some difficulties such as handling and stowing one set of gear and nets, substituting another, all having to be done in the confined space of a trawl deck. Many men are needed to perform this task and in the change over fishing time would be lost.

The 'Delagic' gear, (designed by the UK White Fish Authority to facilitate changing from demersal to pelagic fishing quickly) is aimed at a shoal of fish, so it may be raised quickly and it is for this reason that superkrub otter boards are used.

A large trawler usually carries three or four trawls, one or two fully made up, the others in pieces and depending on circumstances there may be various kinds of trawls as described above.

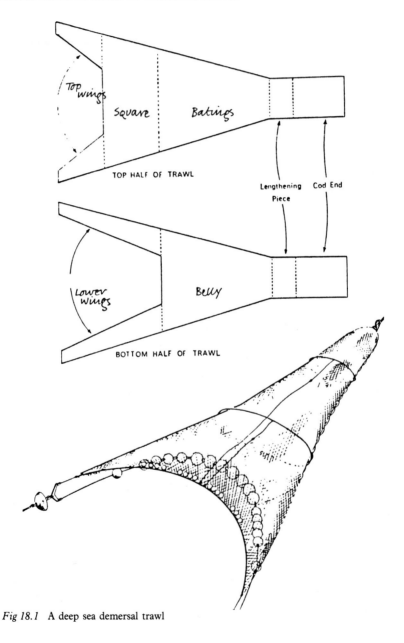

Fig 18.1 A deep sea demersal trawl

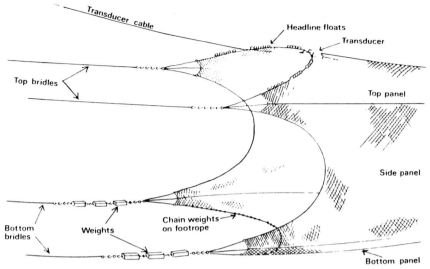

Fig 18.2 Mouth of a pelagic trawl

Side trawling

Shooting the gear

With the gear running clear from the ship, in reasonable weather conditions the engine may be put ahead with the starboard helm so that the doors may be lowered away quickly leading out to starboard.

This should be done after the vessel has gathered a little headway, with the engine stopped and swinging to starboard. The skipper, when sure that the doors and warps are clear, may then adjust his course and speed to be able to pay out the required length of warp. The warps are then put into the towing block by hooking the messenger on to the fore warp, the messenger having previously been led aft from the winch through sheaves near the after gallows. When heaving, the hook will run down the fore warp, it will pick up the after warp and take both warps to the towing block. With the warps safely in the towing block the messenger may be unhooked and warps finally adjusted so that the gear is being towed evenly. The purpose of a towing block is to keep the warp clear of the propeller and thus permit better manoeuvrability.

Manoeuvring when towing

When towing the gear, the skipper or officer of the watch will probably be occupied with keeping the ship on the plot and watching the compass, sounder, and the warps themselves. Nevertheless it should be remembered that the Regulations for Preventing Collisions at Sea must be strictly observed and a good look-out should be maintained at all times. Other vessels may be shooting, hauling or steaming and an adjustment to course or speed made in sufficient time may avert a collision or the fouling of one's gear with that of another ship.

Given the strain on the warps when towing it is obvious a trawler must use greater rudder movements to manoeuvre. It is also more difficult to bring the head round when proceeding into wind as the vessel's ability to pivot around her midship point is restrained.

A side trawler, towing from the starboard quarter carries port helm whilst towing; but will experience no difficulty in altering course to starboard.

Care should be taken when turning that the turn is not made too sharply, as the direction of the lead of the warps will become at too great an angle between the fore and aft line of the ship and the gear is likely to turn over. It will then be necessary to haul it in and valuable fishing time will be lost.

When turning to port a vessel will begin the turn easily for about $30°$ but thereafter will turn more slowly because of the drag at the towing block and pressure of the warp on the starboard quarter.

Should circumstances make it necessary to stop the vessel when towing, stern way will soon be gathered, and a hand should immediately be sent aft to report on the direction of the warps and if they are clear of the propeller before the engine is put ahead again.

Hauling the gear

When ready to haul the gear, speed is reduced according to weather and heading, the warps are knocked out from the towing block and the winchmen are ordered to heave in the warps; first some 5 to 7 fathoms of the fore warp then both together to ensure the doors are lined up. Marks on the warps indicate when the doors are approaching the vessel and the skipper takes the necessary action as to speed and course, with regard to the weather. The trawl doors are brought up into the gallows, hung off on the sling chains, the G links are unclipped and the cables hove in until the dan lenos come into the gallows, when the winch brakes are screwed down and clutches unshipped.

Stern trawling

The basic fishing gear used in stern trawlers is similar to but stronger than that used in side trawlers. The codend, bosoms and bellies are paid out down the ramp when the vessel has a little headway. The doors are connected as in a side trawler and with a little more headway the warps are paid out simultaneously through the stern when the vessel has been brought on to her towing course. Towing blocks are not used. Hauling on a stern trawler is carried out by winching in the warps simultaneously. When the doors are hung up, the dan lenos are brought up to the sheaves, messengers from the winch are made fast to the dan lenos, warps are slackened whilst the messengers take the strain and bring the bobbins and bellies inboard up the ramp. A strop placed around the bellies will serve to haul the codend inboard into a suitable position to allow the fish to run down the hatch into the pounds on the deck below.

Trawling considerations

Towing speed

When a trawl is towed by a powerful stern fisher at the correct speed for maximum catching efficiency, the skipper is well aware that this is close to the point where the gear will lift off the seabed and any error in setting the towing revolutions too high will bring this about. Nevertheless, when towing head to wind the revolutions must be increased in order to cover the ground properly: fear of floating the gear is apt to make one underestimate the increase which is required. A falling off of speed when meeting a head sea produces a distortion of the net.

Hauling speed

For many years it was thought imperative, particularly in stern trawlers, to haul at full towing revolutions in order to avoid damage to the belly of the trawl, which was thought to take place at reduced revolutions. It has since been shown that in the larger vessels the trawl moves across the ground at towing speed when hauling although the revolutions are reduced by 30 to 60 per cent, the ship being practically stopped in the water and the winch hauling in at a four knot rate.

As the trawl is moving through the water at normal towing speed it is not liable to be damaged any more than it would be when towing. This reduction in revolutions allows the trawl to be hauled more quickly, saving time.

It is good practice to heave the first length in by gradually increasing the winch speed from slow at the beginning of hauling to full winch speed by the time it is in, thus avoiding undue strain caused by a sudden jerk which occurs if the winch is put on full power straight away. Similarly it is a wise precaution to ease the ship down two lengths before the last length of warp is paid out when shooting.

Towing hazards

A bottom trawl is prevented from its tendency to be towed up to the surface by being fitted with heavy iron or hard rubber 'Bobbins' attached to the footrope to hold it close to the seabed. These also help to roll over small obstructions. Trawls are nevertheless often snagged on an obstruction or 'fastener' on the seabed, either the footrope becoming caught or one side of the trawl or the doors digging in to a muddy bottom.

Many skippers make use of the Decca Track Plotter to assist them in keeping a record of obstructions on their fishing grounds (but allowance has to be made for the tidal stream effect on gear).

Doors digging in

This type of fastener occurs when the doors dig into a muddy seabed, piling a heap of mud before them and gradually slowing the vessel down until it stops. The warps do not usually pull out in such cases. The trawler must then fall back on the fastened door and hoist it clear. Although the positions of fasteners are always carefully plotted by skippers, this particular hazard, which can also occur on stony ground, is not usually so recorded being of too general a nature and influenced by weight of doors, direction of tow and tidal stream. As this type of fastener almost invariably leads to damage to the belly of the trawl, prompt action in taking way off reduces the extent of the damage and may avoid it altogether.

'One-sider'

Another type of fastener occurs when one side of the trawl is momentarily snagged, at times without pulling the warp out. The result of such a snag is usually the parting of a headline leg or worse still the headline itself. In either case the trawl is no longer fishing and if the headline is parted the longer the tow continues in this state the worse the total damage to the net.

Coming fast – stern trawler

The third case is the unmistakably dangerous fastener where the alarming scream of one or both warps paying out against the brakes is warning enough.

After immediately stopping the vessel, hauling commences, the ship being pulled astern towards the obstruction with the slack warp being taken up at speed until it takes the strain, when the remaining warp is more or less straight up and down. This is the crucial moment when total loss of gear or extreme damage can take place. Watching the strain on each warp can give a useful indication of what has happened down below and allow the best method of clearing the obstacle to be used.

It may be found necessary to heave slowly on one warp, at the same time slacking away on the other in order to pull the gear clear. But no two situations of coming fast are alike and the course of action taken has to suit each individual case, taking weather and tide into account.

It is sometimes possible to clear the trawl from a fastener by slacking away the warp and turning the trawler to head back over the obstruction until the gear comes clear. This technique, however, usually entails accepting fouled gear.

When weather conditions permit, a trawl may sometimes be cleared by awaiting a turn of direction in the tidal stream which will take the trawler back over the fastener and thus bring the gear clear.

Coming fast – side trawler

Should the gear foul an obstruction on the seabed, in a side trawler, the wheel should immediately be put hard over towards the gear and the engine stopped if the ship has not lost headway and the warps are pulling out. When the warps are leading off the side, the engine may be put on dead slow ahead and the helm adjusted so that the ship does not fall on top of the gear. The fore clutch should be shipped as quickly as possible and the warps knocked out from the towing block, so that by heaving on the forward warp, both warps lead away from the ship forward of the beam. The after clutch may then be shipped, and by heaving on both warps and adjusting the helm and engine, the ship may be moved safely over the obstruction with the warps leading up and down. At no time should the warps be allowed to lead under the ship.

If the gear is not cleared when the ship is positioned above the warp obstruction the after warp clutch should be unshipped and the fore warp hove on; the gear should then clear.

If, when heaving in on the fore warp, the ship's head does not come round it may be because the cable or some other part of the gear has been carried away. In such a case, great care must be taken not to foul the rudder or propeller with either of the warps. This may happen if the after gear parts when the outward pull of the door causes the after warp to lead across the ship's stern when being hove in.

Should the ship 'come fast' when towing before strong tides, heavy sea or swell, the above method of hauling should be strictly adhered to. Failure to bring the vessel's head to sea before shipping the after clutch could have disastrous results; with the ship beam on to the seas with warps fastened to the bottom *the resultant loss of stability could be fatal*. If in any doubt the ship should be helped round by heaving on the fore warp whilst paying out on the after warp.

Warnings on coming fast

Upon coming fast it is first necessary to establish that the ship is not a hazard to other vessels and to announce the predicament over the VHF radio. The coming fast signal is sounded on the whistle in daylight, or the appropriate warning lights switched on at night. The skipper's main attention can then be directed aft and the procedure adopted to free the gear from the particular type of fastener.

Parted warp

Once the gear is clear of the obstruction, careful observation will show if it is still attached to both warps. If one cable has parted, tension on the other should indicate that the gear is still attached by the cable. If both seem slack there is reason to think both may be parted.

In both the last two cases an increase in the speed of the vessel over that which is normal when hauling will occur. Stern trawlers are particularly susceptible to locking their doors when clearing a fastener especially with a fresh to strong wind before the beam on the original course.

Lifting and re-settling the gear

Every skipper is conscious of the fact that in some circumstances the gear may be lifted, particularly when making a sharp alteration of course, or bringing the ship round quickly on to a reciprocal course to get back over heavy fish marks which have just petered out. It is always a calculated risk to come round quickly. One method used is to put the vessel on full speed for two or three minutes before altering the wheel and then to apply 20 degrees of helm. For the first 90 degrees of alteration of course, the ship's head comes round fairly slowly until the gear is lifted off the bottom, then the next 90 degrees of change is made much more quickly. By the time that the manoeuvre is completed the log will be showing about six to eight knots. Upon the 180 degree alteration of course the ship's head is kept steady and the propeller revolutions are dropped to towing speed. The gear gradually comes dead astern, still floating, the way will slowly fall off the vessel as shown by the log. At the same time the tension on the warps increases until towing speed and towing tension are resumed. By this time the log speed will indicate if the doors are locked.

If the warp tensions appear normal for the conditions, it may be advisable to drop the towing revolutions for about half a minute to a minute, especially when the new course is before the wind, to ensure that the gear has spread and is settled firmly on the bottom.

The whole exercise is dangerous on muddy grounds where there is a tendency for the door on the inside of the turn to dig in and prevent the lifting of the gear. The possibility of putting a turn in the gear resulting in locked doors is very great.

Foul gear

Despite the ease of shooting from a stern trawler in comparison with the degree of skill required for this operation on a side trawler, it is sometimes possible to shoot a foul gear. This is invariably due to paying away too fast with insufficient tension in the warps, with the result that the whole gear sags and the doors get too close together in midwater. In deep water the angle of the two warps from their sheaves gives a poor idea of the spread of the gear.

Pneumatic winch brakes

The introduction of pneumatic winch brakes has made a great difference to the smoothness of the shooting operation. Manual control is made light and accurate settings of brake pressure can be made. Previously heavy muscular effort was needed and the fierceness of the brake setting depended on the strength of the individual who applied it: two men often being required for the paying away operation. The brakes were reset five minutes after the completion of shooting when the brake drum and linings had cooled down and very often, unknown to the skipper, the last ounce of torque was applied to the brake by using a windlass handle or piece of pipe to extend the lever. This meant a gross oversetting of brake pressure so that only in extreme cases would the warp pull out upon coming fast. Consequently, unexpectedly heavy damage was often found on the trawl after hauling, which could possibly have been avoided by a lighter setting of the brake.

The modern method requires only one man behind the winch when paying away, and a steady brake pressure may easily be set on the completion of shooting. Once time has been allowed for the brakes to cool off, the brake pressure is gradually reduced until the warps just begin to creep out. This pressure is then increased by 5 psi, and that is used as a standard setting for the particular ground in question, having due regard to weather conditions.

Use of load meters

The development in recent years of load meters and their fitting in many of the larger modern stern trawlers has permitted the continuous monitoring of warp tensions, giving warning of fasteners and providing additional information about the warp strain in various situations of trawling. It will be appreciated how much they can assist a skipper when confronted with the situations which have been described above. Quite apart from safety considerations, they have proved of commercial benefit.

Beam trawling

Beam trawling requires a purpose-built vessel. In the North Sea such vessels are usually of Dutch design, approximately 26 metres in length, about 120 GRT and more powerful than normal bottom trawlers of similar size.

The beam trawl may be described as a heavily designed drag net which gives a very effective catch rate for shrimp and flatfish which are to be found on, or just beneath the seabed.

The dimensions of the trawl are matched to the beam size and to the vessel's power.

The modern and larger vessels involved in this technique use metal beams but not in the form of a rigid structure. The rig usually consists of two half beams with shoes attached which are designed to fit into the bore of a central section of metal piping of a slightly larger diameter.

This design allows flexibility of beam length and any operational damage is usually confined to the central pipe section which can be quickly and easily replaced.

The weight of the beam is an important factor, and it is considered better to be too heavy rather than too light.

The beam's length provides the horizontal net opening of the 'mouth' or intake area of the trawl.

The main influence on beam length, in addition to skippers' preferences and fishing regulations, are the vessel's size and horse-power, its gear handling facilities and the area of operation.

The vertical opening of the trawl is achieved by the beam being supported above the seabed by the beam-heads or 'shoes', as they are commonly known. This height is determined by the shoes' dimensions.

As the species being fished are usually found quite close to the seabed the vertical opening of the trawl is never of great height. Consequently, the shoe size may be such as to hold the beam approximately 1½–2 metres above the seabed.

By virtue of their limited height, the beam heads restrict catching depth to a few feet from the seabed, so that the beam trawl is favoured for plaice and sole. In recent years, trawler skippers have utilised more and more heavy tickler chain to 'dig' the fish from the seabed, and some vessels tow as much as 2½ tons of chain hung between the beam heads – impracticable with an otter trawl.

To suit certain fishing conditions, and different species' behaviour, alternative gear arrangements are used. One such variation is the use of chain bellies. This rig is used when working rough stony ground, the large chain meshes allowing any large stones to pass through before reaching the synthetic netting sections. As a result net damage is greatly reduced. Most skippers regard this rigging for rough ground as essential. Chain diameter and weight varies according to vessel and trawl size. In common with otter trawling fewer stones are picked up when towing the gear straight. Changes of direction whilst towing tend to give chain bellies a 'scooping effect'.

Shooting and hauling are simple operations on a twin-rig trawler. Each trawl alternately is hauled alongside and the codend brought aboard by means of a lazy deckie arrangement. Once emptied and re-tied, it is released, and the gear paid away again. When fishing is finished, the booms are topped, the beam trawls lifted clear, and the codends brought onboard.

Beam trawling is generally carried out at considerably greater speeds than normal bottom trawling, 7 to 8 knots being common for the more powerful beamers.

It follows that the hazard experienced on either trawl coming fast is that much greater, particularly as such trawls are generally towed from close amidships.

The effect of one of the trawls coming fast on the seabed, with the weight of the vessel, a powerful engine and possibly a following tidal stream combining to throw weight onto the 'anchored' trawl can readily be appreciated. This can cause the vessel to broach-to and founder unless the load is quickly taken off the overloaded trawl warp. To help in overcoming this danger most beamers have a quick release slip fitted inboard controlling the beam purchase block, which when released shifts the position of tow to gallows in the forepart of the vessel or to bollards aft – transferring the towing strain to a point on the bow or stern so that the vessel may swing safely head or stern on to the gear. (See *Fig 18.3*). Very quick reaction in releasing the slip is nevertheless required on coming fast.

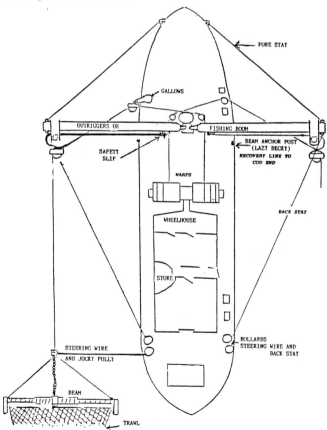

Fig 18.3 Beam trawl rig

19   Seine net fishing

There are two forms of seine net fishing. In principle, the first type is similar to trawling or towing the fishing gear astern, except that it may be considered a more delicate and refined form of fishing compared to the normal stern or side trawling operation of the larger vessel. The second type, known as purse seining, is quite different to tow seining and does not involve the dragging of gear from the stern.

Seining by trawl

Seining by trawl may be carried out in several forms which differ by having various designs of net for use in catching different species of fish. Flat fish are either caught on the bottom or very close to it and a considerable spread of the wings is necessary. Normally the plaice seine does not use a ground rope but may have weights attached to the footrope.

The haddock gear has a greater height of net and may be regarded as the best type for all round seining, being capable of catching higher swimming species as well as flatfish. A ground rope, when used, shorter wings and a very much wider mouth than the plaice seine are features of the haddock seine net.

The deep seine gear, which is used mainly for catching cod, haddock or whiting, requires a greater headline height, which is gained by having a centre rope attached to and running along the middle of the wings and bag to the codend. Apart from this modification, the deep seine is more or less the same as the haddock seine except that it has bridles or sweeps of 10 fathoms or more in length between the wing ends and the dan lenos.

The wing trawl, developed in recent years in order to catch cod, haddock or whiting, has become established as reliable and efficient gear. It is much lighter than the conventional otter trawl but heavier than a similar sized deep seine net. Like the deep seine, sweeps or bridles are used between the wing ends and spreader poles.

Purse seining

Purse seining is a completely different method of fish catching because it does not involve the towing of the fishing gear from astern.

As the name implies, purse seining requires that a net be put over the side in such a manner that it becomes a circular purse or bag to contain the fish.

A buoy is put out from aft to which is attached the headline and the bottom line. The net is mainly rectangular in shape, the upper edge rope floats and the bottom edge rope sinks as the net is paid out. Attached to the bottom edge line by beckets are metal rings through which the purse line is rove. The vessel steams ahead after setting down the buoy, in such a manner as to describe a circle, paying out the net until the circle is completed by the time the buoy has been reached. The result is a net hanging vertically downwards and forming a circular or cylindrical shape. The upper edge lines are brought on board and the purse lines are taken to the pursing gallow and are hove upon until a purse is formed. (See *Fig 19.1*).

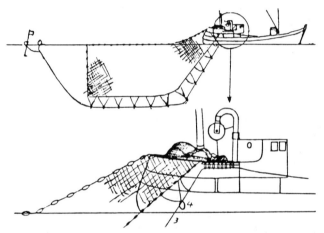

Fig 19.1 Shooting, pursing, hauling and brailing the purse seine. The shooting operation, showing the buoy, the upper edge rope shackled to the buoy as well as the purse line, running through the rings and the net flowing off the upper deck. (*From Modern Fishing Gear of the World 2, published by Fishing News Books Ltd*).

When the purse has been formed and the bottom edge rope is alongside, the upper edge line and net are hauled in until a compact purse has been formed alongside the vessel, with the fish concentrated within. The illustration shows a power block or rollers being used for this operation. The fish may now be brought on board by using a derrick or boom and a basket which is dipped into the net until it is empty.

When closing a purse net the vessel must move to the net not 'bring it to'.

The main hazard experienced in purse seining is that of the vessel fouling her gear through the effects of wind or tide or some form of breakdown in propulsion or winch machinery. Whenever a fouling is suspected the propeller must immediately be stopped (de-clutched). Ideally the safest way out of such fouling is to be towed out astern from the seine net.

Purse seining is also subject to the hazard of fasteners from fouling bottom obstructions. Despite the use of modern echo sounders which permit the skipper accurately to gauge the distance of the seine net from the sea bed it may nevertheless foul an unobserved obstruction, particularly when working close to the bottom. Again the propeller must be stopped immediately and an attempt to clear the net made by working the purse line, heaving one end and releasing the other.

Though it may sometimes be safe for other fishing vessels to pass closer to a purse seiner that is still towing its buoy and has yet to 'make the purse', it should be a rule never to come closer than a quarter of a mile to a vessel that has shot its gear.

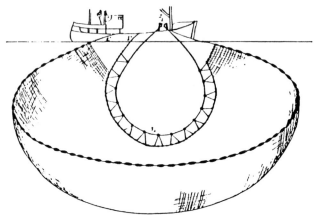

Fig 19.2 The pursing operation showing the retrieving ropes of the wing, the pursing gallow and the purse line. (*From Modern Fishing Gear of the World 2, published by Fishing News Books Ltd*).

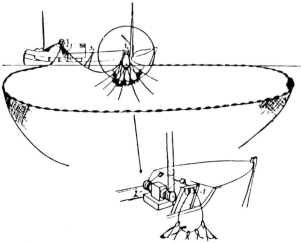

Fig 19.3 The hauling operation, showing the pursing gallow, the pursing winch and the rings unshackled from the purse line and reshackled on a leading line towards the powerblock. (*From Modern Fishing Gear of the World 2, published by Fishing News Books Ltd*).

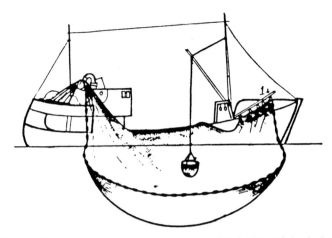

Fig 19.4 The brailing operation, showing the boom on which the breast is hooked and general arrangement of brailing. (*From Modern Fishing Gear of the World 2, published by Fishing News Books Ltd*).

20 Scallop dredging

In British waters scallops are usually caught by towing a dredge across the seabed. The dredge consists of a flat steel frame up to 6 ft wide at the base of which is a toothed cross-bar and a bag to hold the catch. The underside of the bag is made up of small steel rings. The upper side of the bag is made of netting.

Scallop dredging fishing vessels tend to be relatively small, usually around 16.5 metres or less. The smaller boats tow one dredge from each quarter; larger vessels may tow more. The dredges are launched with the vessel proceeding at a good speed for laying out the warps and then towed slowly whilst dredging.

This mode of fishing is therefore also subject to fouling obstructions, in particular telegraph cable, which becomes a major hazard to the vessel's stability if hauled to the surface and higher. *Skippers must therefore be especially careful of not fouling such cables when dredging in areas through which they pass.*

21 Static gear

Whereas the fishing methods discussed so far have involved actively manoeuvring the vessel to catch fish, it should be remembered that the use of static gear is increasing in importance. This technique depends on fish moving to gear set in a particular manner by the vessel and left for some time in one place. The vessel may remain in the vicinity and tend the gear or may leave and only return for periodic checks. The most common static methods are gill nets and longlines, both of which can pose a hazard to and be damaged by 'active' fishing vessels.

Gill nets

The gill net is a large wall of netting which may be set either just above the seabed when fishing for demersal species, or anywhere from midwater to the surface when pelagic fish are being sought.

When working inshore in relatively shallow water, the nets are usually set and anchored in position. An alternative is the drift net which is free to move according to tide and wind conditions.

The top of the net is seized to a float or corkline and the bottom to a leadline. The combined action of the floats and weights maintains the vertical stretch of the net.

With bottom gill nets, sufficient weight is used to keep the leadline on the seabed, while the buoyancy provided by the floats is sufficient only to maintain the vertical stretch. In the case of a midwater gill net on the other hand, sufficient floats are used to overcome the weight of the leadline which is used to maintain the stretch.

Lines from corkline and leadline at each end of the net are connected to lines running from anchors at the seabed to surface buoys which show the location and extent of the gear (and are used in hauling).

A number of gill nets, each several hundred feet in length, may be set end to end in 'fleets', and rather than being set in a straight line may be placed in hooked or curved formations. Usually, however, this is only possible in waters subject to little, if any, tidal movement.

When drift netting a large number of nets will usually be set end to end, extending perhaps for several miles. At the free end of the fleet the net will be secured to a dan buoy; further floats are secured at intervals along the line of set. At the far end, to leeward, the operating vessel secures to the net and drifts under wind action, so ensuring the net remains deployed correctly.

Longlines

The basic method of longlining involves setting out a length of line, which can extend a great distance, to which short lengths of line carrying baited hooks are attached; spaced 5 to 10 feet apart for demersal fish and some 90 to 300 feet apart for pelagic species.

There are wide variations in the dimensions, rigging and operation of the gear depending on the area, species and local tradition, so that only common arrangements and techniques can be described.

Sub-surface longline

With this arrangement the longline is maintained at the desired depth below the surface by regularly spaced lines running up to the surface buoys which carry marker flags.

Bottom longline

The longline lies on or near the bottom and is maintained in position by anchors at each end. The anchors are buoyed and also have marker buoys to show the location of the set, (and to aid retrieval by the operating vessel).

The main set line may be rope or wire, and the gangings may vary from light rope to chain.

Surface (pelagic) longlining

The principle of surface longlining is much the same as those described above. The main line in this arrangement may be 15 to 40 miles in length, with the ganglines between 80 and 100 feet apart.

Large inflatable buoys are placed a considerable distance apart with closely spaced hooks on ganglines. Highflyers with flag, radar reflector, and/or light are placed approximately every mile. Alternatively smaller high density foam floats are spaced more closely; hook spacing is considerably greater. Highflyers are spaced every mile of line. Attendant vessel tends the gear and removes fish as hooked.

It will be appreciated from the description of gill netting and longlining how important it is for other fishing vessels to beware of fouling static gear, particularly at night when many of its floats are not visible.

22 Care of the fish

Each one of the crew must endeavour to ensure that the fish is landed in the best condition. A glance at the voyage analysis will show the difference in price between fish of the best quality and poor fish. A big difference in price generally means that there has been a marked difference in quality. Much of this price difference can be avoided by taking proper care of the fish from the moment it comes on board.

In a side trawler rolling heavily the deck boards should be shipped high across and down the middle of the fish pound to prevent the fish from sliding from one side of the pound to the other. If these boards are only one or two high, the roll sets the fish working or sliding to and fro. Many a good haul of fish has been ruined because this precaution was not taken.

Another reason why the fish should be prevented from sliding about is that the crew have great difficulty in standing among fish on the move and this very much slows down the operation of gutting, thus giving the fish even more time in which to become chafed and soft and eventually unsaleable.

When gutting, only the stomach cavity is cut and not right down into the flesh. Knives should be kept sharp so that the fish is not torn. The flaps of the fish should not be broken, especially haddock, as mutilated fish are practically unsaleable.

The fish pound must be well hosed down before the codend is emptied on to the deck. Fish dropped on deck among the guts and offal of the previous haul are immediately contaminated and have taken the first step to the fish meal factory.

The washing of fish is a most important feature. A plentiful supply of water entering the washer at considerable force is essential in order to avoid the entry of dirty fish to the fishroom. The boatswain must ensure that the water supply is sufficient at all times and that only well washed fish go down the hatch.

Before stowing any fish, the fishroom pounds should be washed out again to ensure their thorough cleanliness. Care should be taken to remove all the slime from any surface liable to come in contact with the fish, especially above and below the battens. The slush wells must be disinfected with chloride of lime or some other recommended disinfectant, the odour of which cannot be absorbed by the fish, and be thoroughly cleared of any small pieces of wood, brush bristles or any matter that may stop up the holes in the rose or strum boxes. The space under the fishroom floor must be kept sweet and clean by pouring some buckets of disinfectant, similar to that used in the slushwell, down the deck pumps on the outward passage and allowing the mixture to swill about under the floor helped by a certain amount of clean sea water. The whole area should be pumped dry before taking any fish below. Slush wells are also found in ships with double-bottom tanks.

Each fishroom pound is given a good 'bottom' of ice, at least one board, *ie* about nine inches deep, just before taking the fish into it. The ice will not then solidify. If the bottoms are made too soon they become very hard and the first layer of fish in the pound becomes badly squashed.

Plenty of ice must be distributed evenly among the fish and along the woodwork and against the ship's side to ensure that the fish does not come in contact with the wood, which must always be suspected of harbouring bacteria. Always look after the first caught fish, as if the trip was going to be of long duration no matter how promising the prospect of a short trip may appear.

Never go too far above the battens before putting the shelves on or squashed fish will be the result. Never miss a shelf no matter how hard pressed one appears to be for time.

Avoid having more than one fishroom hatch off at a time, especially during the summer months when the temperature of the outside air is considerably higher than that of the fishroom.

Flatfish have a special tendency to slide about when they have been put down in the fishroom. This starts the formation of a frothy slime, which causes the fish to slide even more and the whole pound of fish could be ruined. To prevent this the pound should be halved by shipping up the fore and aft boards dividing the 'front half' from the 'back half' and putting a good layer of ice on top of the fish.

Freezer trawlers

The freezing of fish is not a magical process that turns bad fish into good. The same care must be used in dealing with fish as is used in fresh fish trawlers; that is they must be properly gutted and well washed. The fish should then be taken to the freezers as quickly as possible and be frozen. A general rule is that fish should be frozen for about four hours, but this can vary a little depending on the type of freezer, the surrounding or 'ambient', temperature and the oil content of the fish.

Freezers commonly used for whole fish are called vertical plate freezers, as opposed to horizontal plate freezers which are used for fillets.

A very important aspect of freezing is that before the fish go into the plate freezers they should be sorted for size and type. Once the block of fish is frozen it is impossible to separate the fish from the block. If the fish are mixed, the frozen block is an unsaleable product and causes loss.

Once frozen, the blocks of sized and typed fish must be carefully labelled. This assists in the proper sorting of the fish during the landing operation. Without correct labelling it is almost impossible for the people landing the fish to distinguish one type from another.

A second important aspect in a freezer trawler is that, when the blocks are put into the fish room, they should be properly stowed and boarded up. Failure to do this may allow the blocks to slide about, not only damaging the fish and parts of the fish room and fittings and possibly causing injury to personnel, but causing a stability problem.

Refrigerated sea water

Refrigerated sea water (RSW) offers an efficient means of rapid chilling, either as a temporary measure prior to processing or as an alternative to ice or freezing when stowing the catch onboard. RSW is mainly used for holding herring, mackerel or other pelagic species in bulk tanks. Some species such as cod cannot be kept long in RSW without absorbing salt.

The chilling of the brine may be achieved in a heat exchanger or by means of coils in the tank.

The RSW system of a tuna clipper is somewhat different. The fish are stowed in a series of tanks in which cooling coils bring the temperature down rapidly. When a certain temperature level is reached, the brine is drained and the temperature pulled down to a minus 18 deg C 'dry'.

Trials have been made with portable 'modular' insulated tanks, which are loaded on a boat empty and filled with herring, sea water and ice as fishing progresses. The fish remain in these containers until delivered by road or rail to their destination.

Inshore fish

Care of inshore fish.   There is little difference in initial quality between fish caught on inshore grounds and fish of the same species and condition caught in deeper waters. There is little justification for the belief that inshore fish will not keep well when properly looked after. These fish, however, are often of small size and full of food, and will spoil more quickly than fish with empty stomachs.

Fish without ice may look very attractive when landed soon after catching, and may taste good if eaten right away; but if they have to travel long distances overland these characteristics are quickly lost if they are not gutted and iced as soon as they are caught.

If fish have to lie on deck for some time, protect them with an awning or cover of some kind; don't leave them exposed on the open deck, particularly in the sun. Keep them moist in pounds. Wash the gutted fish to remove blood and offal; rinse the large ones by hand and pay particular attention to the belly; small fish can be washed in open-mesh baskets or in a tank of running water.

Part V – Safety and survival at sea

23 Danger messages

The International Convention for the Safety of Life at Sea (SOLAS) 1974 requires that the master of every ship which meets with dangerous ice, a dangerous derelict, or any other direct danger to navigation, or a tropical storm, or encounters sub-freezing air temperatures associated with gale force winds causing severe ice accretion on superstructures, or winds of force 10 or above on the Beaufort scale for which no storm warning has been received, is bound to communicate the information by all the means at his disposal to ships in the vicinity, and also to the competent authorities at the first point on the coast with which he can communicate. The form in which the information is sent is not obligatory. It may be transmitted either in plain language (preferably English) or by means of the International Code of Signals. It should be broadcast to all ships in the vicinity and sent to the first point on the coast to which communication can be made, with a request that it be transmitted to the appropriate authorities.

The transmission of messages respecting the dangers specified is free of cost to the ships concerned.

All radio messages issued under this Regulation shall be preceded by the Safety Signal, as described later in this chapter.

The following information is required in danger messages:

Ice, derelicts and other direct dangers to navigation

The kind of ice, derelict or danger observed; its position time and date (GMT) when last observed.

Tropical storms (Hurricanes in the West Indies, typhoons in the China Sea, cyclones in Indian waters, and storms of a similar nature in other regions)

A statement that a tropical storm has been encountered. This obligation should be interpreted in a broad spirit, and information transmitted whenever the master has good reason to believe that a tropical storm is developing or exists in his neighbourhood.

Time, date (Greenwich Mean Time) and position of ship when the observation was taken.

As much of the following information as is practicable should be included in the message:

— barometric pressure, preferably corrected (stating millibars, millimetres, or inches, and whether corrected or uncorrected);
— barometric tendency (the change in barometric pressure during the past three hours);
— true wind direction;
— wind force (Beaufort scale);
— state of the sea (smooth, moderate, rough, high);
— swell (slight, moderate, heavy) and the true direction from which it comes. Period or length of swell (short, average, long) would also be of value;
— true course and speed of ship.

Subsequent observations

When a master has reported a tropical or other dangerous storm, it is desirable, but not obligatory, that further observations be made and transmitted hourly, if practicable, but in any case at intervals of not more than three hours, so long as the ship remains under the influence of the storm.

Winds of force 10 or above on the Beaufort scale for which no storm warning has been received.

This is intended to deal with storms other than the tropical storms referred to above; when such a storm is encountered, the message could contain similar information to that listed under that paragraph but excluding all the details concerning sea and swell.

Sub-freezing air temperatures associated with gale force winds causing severe ice accretion on superstructures

— Time and date (Greenwich Mean Time).
— Air temperature.
— Sea temperature (if practicable).
— Wind force and direction.

24 Distress and rescue procedures

We have already noted under the International Regulations for Preventing Collisions at Sea, 1972, Annex IV the signals which, used or exhibited either together or separately, indicate distress and need of assistance.

Obligations and responsibilities

The International Convention for the Safety of Life at Sea (1974) states that a master or skipper of a ship at sea receiving a signal from any source that a ship or aircraft is in distress is bound to proceed with all speed to the assistance of the persons in distress, informing them if possible that he is doing so. If he is unable, or in the special circumstances of the case, considers it unnecessary to proceed to their assistance, he must enter in the log book the reason for failing to proceed to the assistance of those in distress.

The master of a ship in distress, after consultation, so far as may be possible, with the masters of the ships which answer his call for assistance, has the right to requisition such one or more of those ships as he considers best able to render assistance. It shall be the duty of the master or masters of the ship or ships requisitioned to comply with the requisition by continuing to proceed with all speed to the assistance of those in distress.

The master or skipper shall be released from the obligation imposed in the first paragraph, when he learns that one or more ships other than his own have been requisitioned and are complying with the requisition.

These are briefly the obligations and duties imposed on those in charge of ships at sea by the International Convention for the Safety of Life at Sea (1974). Skippers of fishing vessels on seeing or hearing any of the signals indicating distress and need of assistance should take the appropriate action if it is possible to do so. A full description of search and rescue procedures is given in the IMO publication *Merchant Ship Search and Rescue Manual*. (MERSAR).

General arrangements for search and rescue (SAR)

In general, distress incidents fall into two main categories:

— Coastal – in which some or all of the following may be available to assist: ships, aircraft, helicopters and shore-based life-saving facilities.

— Ocean – in which ships and long range aircraft may be available and in the more remote areas, only ships may be available.

To supplement the efforts of the ships in the area of a vessel in distress there is an extensive international SAR organisation. The SAR measures vary from country to country but most maritime nations have coast radio stations which always play an important part by guarding the international distress frequencies.

Search and rescue (SAR) areas

Designated SAR areas are under the control of Headquarters coordinating the facilities made available, both ships and aircraft. They are linked to coast radio stations and adjacent SAR areas and have access to satellite communications.

Distress frequencies and procedures

A ship in distress should always transmit the appropriate alarm signal on the international distress frequencies 500 kHz or 2182kHz followed by the distress message. It may be transmitted on any frequency available on which attention might be attracted, such as an inter-ship frequency which may be in use in local areas. Before changing frequency however, adequate time should be allowed for reply.

In certain cases, depending on location, it may be helpful to transmit the distress call and message on VHF Channel 16 (156.8 MHz), eg in estuaries, when close to coast radio stations, or when in company of other ships where it is known that VHF watch is kept.

The radio watch on the international distress frequencies which ships must keep when at sea is one of the most important factors in the arrangements for the rescue of those in distress at sea. Since these arrangements must often fail unless it is possible for ships to alert each other or to alert or be alerted by shore stations for distress action, every ship should make its contribution to safety by keeping watch on one or other of these distress frequencies for as long as practicable whether or not required by regulation.

Coast radio stations play an important part in rescue work by maintaining constant watch on the distress frequencies so that in the event of a distress signal being heard, they can not only alert other ships in the vicinity but also the proper shore authorities, who can bring the SAR services into action as required.

In the United Kingdom all coast radio stations keep continuous watch on 2182 kHz and Channel 16 VHF (and most on 500 kHz also) for distress, urgency and safety calls. They are linked to the Coastguard organisation.

HM Coastguard, who are responsible for UK civil maritime SAR areas also maintain continuous radio listening watch on 2182 kHz and Channel

16 VHF, (and visual watch where appropriate).

All fishing vessels of 12 metres or more in length registered in the UK are required to carry radio telephony equipment and if they proceed outside the area prescribed in Fishing Vessels Radio Rules they have to carry radio telegraphy. It is also a requirement that a continuous watch be kept; this is usually carried out by having the receiver switched through to a loudspeaker in the wheelhouse, so that the watchkeepers may listen on the distress frequency when the radio officer is off duty.

Lifeboats

Most lifeboats are fitted with VHF radio and MF R/T. Many have radar and D/F equipment covering 2182 kHz.

It is imperative to give an accurate position when sending a distress or urgency message. Not only will the lifeboat reach the distress scene more quickly, but the steaming range will be reduced if extra ground is covered by having to look for a casualty which has been given an inaccurate position.

Inmarsat maritime satellite service

A Maritime Satellite Service is available throughout most of the Atlantic, Pacific and Indian Oceans but not in certain parts of the polar regions nor in a zone in the Pacific Ocean along a narrow longitudinal strip off the west coast of South America at about 104°W. Communications are established via satellites in geostationary orbits, positioned centrally over each of the three ocean areas.

Ships equipped with INMARSAT equipment may communicate with shore telephone and telex subscribers or other ships via Coast Earth Stations and a satellite link. Calls between equipped ships are routed via 2 satellite links and one or two coast earth stations depending on the location of the ships.

Telephone services provided through coast earth stations are either in an automatic or operator-assisted mode. In the automatic mode calls may be made direct to subscribers provided the appropriate facilities exist between the coast earth station and the country of destination of the call. If these facilities are not available then operator assistance is required.

Distress, urgency and safety services provided include distress priority telex and telephone communications, medical advice and medical and maritime assistance services. Distress calls to coast earth stations are normally automatically routed to the associated Rescue Coordination Centre.

Details are contained in Admiralty List of Radio Signals Vol I.

Procedure when sending a radio distress or urgency signal

All British registered fishing vessels and those of most maritime nations whether they are fitted with radio telegraphy or radio telephony are fitted with an automatic alarm signalling device.

The radio telegraph alarm signal, which consists of 12 dashes sent in one minute by an automatic keying device, activates auto-alarms on other ships within range and it should be followed by the radio telegraph distress signal which consists of: the distress signal, SOS, sent three times; then word DE followed by the call sign of the ship sent three times. If circumstances permit, an interval of two minutes should be allowed to elapse to enable operators to reach their apparatus and bring it into operation before sending the distress message.

The radio telephone alarm signal is also intended to give preliminary warning to other ships and coast stations either aurally or by activation of radio telephone auto-alarms. It consists of two tones transmitted alternately over a period of at least 30 seconds. The radio telephone distress signal consists of the word 'Mayday' spoken three times, followed by the words 'This is' followed by the name of the ship, repeated three times.

On hearing either of these two distress signals being transmitted by W/T or R/T all stations must cease all transmissions capable of interfering with the distress call or associated messages, and must listen on the frequency used for the distress call.

Distress calls should not be addressed to a particular station but should be repeated with discretion until the distress message can be sent. The distress message consists of the distress call, followed by the name of the ship in distress and information concerning her position, the nature of the distress and the kind of assistance required. It is important that the position of the vessel be given in terms of latitude and longitude or if aground by giving a clear geographical indication, *ie* two miles north of Flamborough Head. All relevant information such as weather, abandonment, *etc* should also be included in the distress message.

In radio telephone ships, cards of instructions giving a clear summary of radio telephone distress procedure must be displayed in full view of the R/T operating position.

The form of cards to be displayed are shown in Appendix 4. The words printed in bold type should be printed in red.

Urgency signals

The radio telegraphy urgency signal $(-\cdot\cdot-/-\cdot\cdot-/-\cdot\cdot-)$ repeated three times and the radio telephone urgency signal, which consists of the group of words 'Pan Pan' repeated three times, are provided for use in cases in which a ship making a call has a very urgent message to transmit concerning the safety of the ship or some person on board or within sight but it does not necessarily imply that the ship is in imminent danger or requires immediate assistance. The call has immediate priority over all other communications except distress calls and it should be used in all cases in which the sending out of an SOS and Mayday signal is not justified. The urgency signal and message must, where practicable, be addressed to a specific station, *ie* to the nearest coast radio station or to another ship known to be in the vicinity. The urgency signal may also be used when the master or skipper of a ship desires to issue a warning that circumstances are such that it may become necessary for him to send out a distress signal at a later stage but in such a case the signal need not be sent to a specific station. If precautionary action ceases to be necessary, the message should be cancelled at once.

Safety signal

The spoken word 'Securite' (pronounced say-cure-e-tay) repeated three times indicates that the station is about to transmit a message concerning the safety of navigation or giving important meteorological warnings.

If you hear these words, pay attention to the message, acknowledge and call the skipper:

Mayday (Distress)	Indicates that a ship or aircraft is threatened by immediate danger and requests immediate assistance.
Pan Pan (Urgency)	Indicates that the calling station has a very urgent message to transmit concerning the safety of a ship, aircraft or of a person.
Securite (Safety)	Indicates that the station is about to transmit a message concerning the safety of navigation or giving important meteorological warnings.

Silence periods

On the distress frequencies of 500 kHz, 2182 kHz and VHF Channel 16 the following silence periods have been set aside during which times only distress, urgency and safety calls should be made. The silence periods leave the frequencies clear for emergency transmissions and vessels on passage and fishing should observe strict radio silence at these times. Ships in distress, survivors in liferafts, *etc* should, if possible, use these silence periods to transmit their calls and so make sure that they have been heard. Silence periods are of three minutes duration and begin:

Radio telegraphy silence period:	At 15 mins and 45 mins past each hour, GMT.
Radio telephony and VHF *silence period:*	*At each hour and half hour, GMT*

Direction finding and homing

Subsequent to the transmission of the distress message sent on 500 kHz

(W/T), two dashes of ten to fifteen seconds duration should be transmitted followed by the call sign to enable coast radio stations and other vessels to take D/F bearings. This transmission to be repeated at regular intervals.

In cases where 2182 kHz (R/T) is used, similar action should be taken, using a continued repetition of the call sign or name of ship or a long numerical count in place of the two long dashes mentioned in the previous paragraph.

Portable radio equipment which may be carried on fishing vessels for use in liferafts and capable of transmission and reception on 2182 kHz should be used by survivors in liferafts, preferably during the silence periods so that SAR craft can home on the signal.

If circumstances change and assistance is no longer required distress and urgency messages should always be cancelled as soon as possible.

Action by assisting ships

A ship may receive by radio an alarm and/or distress signal from a ship or aircraft in distress, either directly or by relay, by a signal emitted by an emergency position indicator radio beacon (EPIRB), or by a visual or sound signal.

The immediate action to be taken on receipt of the message, whether received by radio, visually or by sound, would be to acknowledge receipt and if appropriate re-transmit the distress message. If the distressed vessel is seen or heard – proceed to the vessel. If the signal has been received by radio – try to take D/F bearings of the distress message, give your identity, position and ETA. Maintain a continuous watch on the distress frequency in use, operate the radar and, if in the vicinity of the distress, post extra lookouts and maintain VHF watch on channel 16. Repeat the distress call on both frequencies, 500 kHz and 2182 kHz, if you are able to for the benefit of other stations.

While proceeding to the area, plot the position, course and speed of any other assisting vessels. If it appears that you may be one of the first ships on the scene, try to construct an accurate picture of the circumstances attending the casualty. The important information needed is weather, wind, sea, swell, visibility, *etc*, time of abandonment, number of crew on board, injuries, number of survival craft *etc*.

When on route to the distressed vessel prepare insofar as you are able some of the following which may be considered necessary: towing gear, line throwing apparatus, ships side lines, scramble nets, heaving lines, rope ladders, liferaft, derrick and runner, and any medical preparations which survivors may need.

Assistance by aircraft

Military fixed wing aircraft used on SAR duties usually carry droppable survival equipment and pyrotechnics. These aircraft may be able to assist a ship in distress by:

— Locating her when her position is in doubt and informing the authorities so that ships in the vicinity going to her assistance may be given her precise position.
— Guiding surface craft to her position or, if the ship has been abandoned, to survivors in lifeboats, liferafts or in the sea.
— Keeping the casualty under observation.
— Marking a position by marine marker, smoke float, or flame float and illuminating an area to assist rescue operations.
— Dropping survival equipment.

The air-dropped survival equipment carried by many military SAR aircraft usually comprises a series of containers carrying a liferaft (which inflates automatically on striking the water), and the other supplies, usually connected by line to aid recovery.

Additional liferafts can also be dropped. The contents should be secured to avoid loss in rough seas and not unpacked until required.

Outer containers should be discarded.

Aircraft search areas are established on latest estimates of a casualty's position. At night they will search an area at between 3,000 and 5,000 feet, or below cloud firing a green Very cartridge about every five minutes. On seeing a green flare the following action must be taken:

— Wait for the glare of the green flare to die out.
— Fire one *red* flare.
— Fire another red flare after about 20 seconds (this enables the aircraft to line up on your bearing).
— Fire a third red flare when the aircraft is overhead, or if it appears to be going badly off course.

Failure to do so could lead to aircraft moving to another search area.

Points to note:

— Each liferaft/lifeboat must carry at least three red flares.
— If the aircraft has been diverted to the search from another task it may fire flares of another colour (except red) – reply as above.
— If all else fails, use any means at your disposal to attract attention.

Assistance by helicopter

In effecting any transfers by helicopter the following procedures should be followed:

— The ship must be on a steady course giving minimum ship motion with the relative wind from about $30°$ on the bow. If this is not possible the ship should remain stationary head to wind or follow the instructions from the helicopter crew.

— Relative wind should be indicated (by flag or other means). It is desirable that it be at least two points on the bow especially if making smoke.

— Clear as large an area of deck and rigging as possible.

— All loose articles must be securely lashed down or removed from the transfer area. The downwash from the helicopter's rotor will easily lift unsecured covers, tarpaulins, hoses, rope, gash, *etc*, thereby presenting a severe flying hazard. Even pieces of paper in a helicopter engine can cause a helicopter to crash.

— On no account must the helicopter winch wire or load be allowed to foul any part of the ship or rigging. In the event of a load or the winch wire becoming snagged, the helicopter crew will cut the wire.

— *The winch wire should be handled only by personnel wearing rubber gloves.* A helicopter can build up a charge of static electricity which, if discharged through a person handling the winch wire, can kill or cause severe injury. The helicopter crew will normally discharge the static electricity by dipping the winch wire into the sea, or by allowing the hook to touch the ship's deck, prior to beginning the operation.

— The helicopter will approach the ship from astern heading into the relative wind and can lower on to or lift from the clear area (or boat towed astern). The helicopter's height will be conditioned by the length if its winch wire and ship's obstructions.

— When landed from the helicopter beware of rotating tail rotor.

Use of helicopters, rescue, medical evacuations

Whenever possible, and if time allows, all the preceding safety precautions should be taken. However, in a distress situation it may not be possible to meet all the requirements. Under such circumstances the operation may necessarily be slower, but it should be borne in mind that helicopters have operational limitations and should not be delayed at the scene of the rescue. Cases have arisen where the rescue has been hampered by survivors trying to take personal belongings with them. In distress situations, transfers are limited to personnel only.

Once the helicopter has become airborne, the speed with which it locates the ship and the effectiveness of its work depends to a large extent on the co-operation of the ship itself. From the air, especially if there is a lot of shipping in the area, it is very difficult for a helicopter pilot to find the particular ship he is looking for, unless that ship uses a distinctive distress signal which can be clearly seen. The orange coloured smoke signal carried in liferafts, a well-trained Aldis lamp and the use of the heliograph in bright sunlight could be used. The use of these signals will save valuable time in locating the casualty.

It is essential that the ship's position be given accurately if the original distress call has been made by radio, along with information on type of ship and colour of hull, if time allows.

Helicopters are well practised in rescuing survivors from either the deck or the sea and two methods are employed:

—The survivor, whether on deck or in the water is rescued by means of a strop. Whenever possible a crewman is lowered from the helicopter together with a strop which is secured around the survivor's back and chest, and both are winched up. On occasions it may be necessary for the survivor to position the strop himself and give the thumbs up sign when ready to be hoisted, subsequently keeping his arms tucked into his sides.

—If a survivor is injured to the extent that a strop cannot be used a crewman is lowered with a stretcher from the helicopter. The injured man can be secured in the stretcher and hoisted up. If he is already secured in a Neil Robertson type strecher, this can be lifted into the helicopter or placed in the rigid framed stretcher belonging to the helicopter.

Designated SAR helicopters can communicate with lifeboats on VHF on the marine distress band 156.8 MHz (Channel 16). They normally have no MF communication equipment.

Wave-quelling oils

There is little doubt that oils, when properly used, are very effective wave quelling agents. However, when survivors are likely to be in the water, the pumping of oil should only be carried out when absolutely necessary and then the greatest of care should be taken.

Experience has shown that vegetable and animal oils, including fish oils, are the most suitable oils to use as quelling agents and are the least harmful to men in the water. If none of the former oils are available, lubricating oils should be used. Fuel oil should never be used unless absolutely unavoidable and then only in small quantities. Tests have shown that 200 litre of lubricating oil discharged slowly to leeward, through a rubber hose with an outlet just above the sea, while the ship proceeds at slow speed can be an effective agent for quelling seas over an area of at least 4,500 square metres.

Emergency Position Indicating Radio Beacons

When disaster occurs in a vessel it is not always possible to send a distress message, nor is there always the possibility to collect survival radio equipment before abandoning ship.

An Emergency Position Indicating Radio Beacon (ERIRB) is a device designed to transmit a distress signal automatically when activated. One model floats off automatically from a vessel which has sunk or capsized. On being immersed to a certain depth it is ejected by a hydro-static arrangement, immediately activating the transmitter. Some models are portable and also have a manual activating system (enabling them to be taken into a liferaft): others are already integral to a liferaft and are automatically activated when it is inflated. Yet other 'pocket sized' hand-held beacons are activated by the operator. There is also a type of EPIRB housed in a tethered buoy which remains attached to a sunken vessel.

On activation, these beacons at present transmit on 121.5/243 MHZ distress frequencies which are monitored by aircraft. SAR aircraft and ships can then home on to the beacon directly by DF.

The introduction of the COSPAS/SARSAT (Search and Rescue Satellite Aided Tracking System) is greatly extending SAR capability.

COSPAS/SARSAT has placed equipment on four satellites to receive distress alert transmissions from EPIRBs and relay those transmissions to a network of ground stations. Unlike the more random detection by aircraft of transmissions from beacons, the COSPAS/SARSAT system is both methodical and reliable; and the satellites, each of which orbits the earth every 100 minutes, provide coverage of the whole globe. Global coverage is achieved in a mean time of 2 hours, less nearer the polar regions.

On picking up an EPIRB transmission the satellite re-transmits it to ground stations called Local User Terminals –LUT's – in the course of its orbit.

The LUT detects the distress signals and usually well within 20 minutes of the end of a satellite pass, it processes the signals to determine the location of each beacon. It then sends alert messages to a Rescue Co-ordination Centre.

The system, operating on 406 MHZ, is now in use. It provides position accuracy of the order of 5 n.m. and is capable of identifying the craft from which a distress transmission is sent, thereby improving the suitability of the search and rescue service to be provided. Satellite installation has been designed so that transmissions can be stored and processed onboard for relaying to ground stations if none is in view when the alert is transmitted.

The COSPAS/SARSAT system also responds to beacons transmitting on 121.5/243 MHZ if the distress and ground station are both visible to the satellite. This arrangement presently takes in most of the North Atlantic, the seaboards of the United States, Canada and much of Central America, the Mediterranean and North European waters. Position accuracy on 121.5/243 MHZ beacon transmissions is of the order 15 – 20 n.m.

With the progress of the COSPAS/SARSAT programme the use of EPIRBs is becoming increasingly widespread. It is mandatory in the fishing fleets of a number of nations, the free float hydrostatic release model being mostly favoured, or that activated by liferaft inflation.

218

The International Maritime Organisation (IMO) in making its decision about the type of EPIRB to be used in the Future Global Maritime Distress and Safety System (FGMDSS), due to come into operation in the 1990s, has decided that the 406 MHZ frequency satellite EPIRB system should become the internationally accepted system.

The British government already supports the use of 406 MHZ beacons and UK fishing vessels over 12 metre are to be required to carry 406 MHZ EPIRBs.

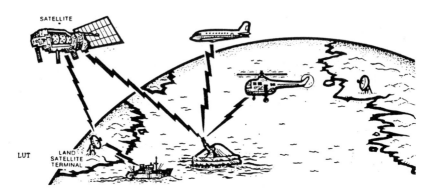

Fig 24.1   COSPAS/SARSAT system (courtesy EMTRAD LTD, suppliers of LOCAT beacons)

25 Visual signals, ships in distress, shore stations and aircraft

In accordance with the Convention for the Safety of Life at Sea 1974, contracting Governments undertake to ensure any necessary arrangements are made for coast watching and for the rescue of persons in distress at sea around their coasts.

— *Replies from life saving stations or maritime rescue units to distress signals made by ship or person:*

Signals	Signification
By day – Orange smoke signal or combined light and sound signal (thunderlight) consisting of three single signals which are fired at intervals of approximately one minute.	'You are seen, assistance will be given as soon as possible.' (May be repeated.)
By night – White star rocket consisting of three single signals which are fired at intervals of approximately one minute.	

If necessary the day signals may be given at night or the night signals by day.

— *Landing signals for the guidance of small boats with crews or persons in distress:*

Signals	Signification
By day – Vertical motion of a white flag or the arms or firing of a green star signal or the code letter 'K' (— · —) given by light or sound signal apparatus.	'This is the best place to land'.
By night – Vertical motion of a white light or	

220

flare or firing of a green star signal or the code letter 'K' (— · —) given by light or sound signal apparatus. A range (indication of direction) may be given by placing a steady white light or flare at a lower level and in line with the observer.

By day – Horizontal motion of a white flag or arms extended horizontally or firing of a red star signal or the code letter 'S' (· · ·) given by light or sound signal apparatus.

'Landing here highly dangerous.'

By night – Horizontal motion of a white light or flare or firing of a red star signal or the code letter 'S' (· · ·) given by light or sound signalling apparatus.

By day – Horizontal motion of a white flag, followed by the placing of the white flag in the ground and the carrying of another white flag in the direction to be indicated or firing of a red star signal vertically and a white star signal in the direction towards the better landing place or signalling the code letter 'S' (· · ·) followed by the code letter 'R' (· — ·) if a better landing place for the craft in distress is located more to the right in the direction of approach or signalling the code letter 'L' (· — · ·) if a better landing place for the craft in distress is located more to the left in the direction of approach.

'Landing here highly dangerous. A more favourable location for landing is in the direction indicated.'

By night – Horizontal motion of a white light or flare, followed by the placing of the white light or flare on the ground and the carrying of another white light or flare in the direction to be indicated. The use of a red star signal and visual/sound signals as described above to be used as appropriate.

— *Signals used by aircraft engaged on search and rescue operations to direct ships towards an aircraft, ship or person in distress:*

The following procedures performed in sequence by an aircraft mean that the aircraft is directing a surface craft towards an aircraft or a surface craft in distress:

— circling the surface craft at least once;
— crossing the projected course of the surface craft close ahead at a low altitude, opening and closing the throttle or changing the propeller pitch;
— heading in the direction in which the surface craft is to be directed.

Repetition of such procedures has the same meaning.

The following procedure performed by an aircraft means that the assistance of the surface craft to which the signal is directed is no longer required:

— crossing the wake of the surface craft close astern at a low altitude, opening and closing the throttle or changing the propeller pitch.

Note: Advance notification of changes in these signals will be given by the national administration.

Breeches buoy

Co-operation between a ship's crew and Coastguard or competent authority in the use of rocket rescue equipment

Should lives be in danger and the endangered vessel is in a position where rescue by rocket equipment is possible, a rocket with line attached will be fired from the shore above the vessel. Take hold of this line as soon as possible and, having done so, signal to the shore as indicated in A(*1*) below.

Alternatively, should the vessel carry a line-throwing appliance and this is first used to fire a line ashore, this line will not be of sufficient strength to haul out the heavier line and those on shore will, therefore, secure it to a stouter line. When this is done, they will signal as indicated in A(*1*) below. On their signal, the line should be hauled in until the stouter line is on board. The rocket line is a very light line and should be hand hauled back on board; the winch should not be used because any sort of check or snag could part the line and time would be lost.

If a rocket line is received from ashore, make the appropriate signal to the shore when it is held and proceed as follows:

— When the appropriate signal, *ie* 'haul away', is seen from the shore, haul on the rocket line until the tail block with an endless fall rove through it (called the 'whip') is aboard.

— Make the tail block fast close up to the mast or other convenient position, bearing in mind that the fall should be kept clear from chafing any part of the vessel and that space must be left above the block for the hawser. Unbend the rocket line from the whip. When the tail block is made fast and the rocket line unbent from the whip, signal to the shore again as below A(*1*).

— As soon as this signal is seen on the shore a hawser will be bent on to the whip and will be hauled off to the ship by those on shore. Except when there are rocks, piles or other obstructions between ship and shore, a bowline will have been made with the end of the hawser round the hauling part of the whip.

— When the hawser is on board, the bowline should be cast off. Then having seen that the end of the hawser is clear of the whip, the end should be brought up between the two parts of the whip and made fast to the same part of the ship as the tail block, but just above it and with the tally board close up to the position to which the end of the hawser is secured (this will allow the breeches buoy to come right out and will facilitate entry to the buoy).

— When the hawser has been made fast on board, unbend the whip from the hawser and see that the bight of the whip has not been hitched to any part of the vessel and that it runs free in the block. Then signal to the shore as A(*1*) below.

— Those on shore will then tighten the hawser and, by means of the whip, haul the breeches buoy out to the ship. The person to be rescued should get into the breeches buoy and sit well down. When secure he should signal again to the shore as indicated below in A(*1*) and those ashore will haul him to land.

— During the course of the operation, should it be necessary to signal either from the ship to the shore or from shore to shore to ship to 'Slack away' or 'Avast hauling', this should be done as in paragraph A(*2*).

It may sometimes happen that the state of the weather and the condition of the ship will not allow a hawser to be set up; in such circumstances a breeches buoy will be hauled off by the whip, which will be used without a hawser.

The system of signalling must be strictly followed. It should, however, be noted that the rescue operations as a whole will be greatly facilitated if signal communication (by flashing lamp or VHF) is established between the ship and shore or lifeboat. Coastguard rescue companies very often have trained signalmen.

A. Signals to be employed in connection with the use of shore life-saving equipment.

Signals	*Signification*
(1) *By day* – Vertical motion of a white flag or the arms.	In general 'Affirmative'. Specifically 'Rocket line is held'.
By night – Vertical motion of a white light or flare or a green star.	'Tail block is made fast' 'Hawser is made fast' 'Man is in breeches buoy' 'Haul away'
(2) *By day* – Horizontal motion of a white flag or arms extended horizontally.	In general 'Negative'. Specifically 'Back away'. 'Avast Hauling'
By night – Horizontal motion of a white light or flare or a red star.	

B. Signals to be used to warn a ship which is running into danger.

Signals	*Signification*
The International Code signals 'U' or 'NF'. The letter U ($\cdot \cdot -$) flashed by lamp or made by fog horn, whistle, *etc.*	'You are running into danger'

If it should prove necessary, the attention of the vessel is drawn to these signals by a white flare, a rocket showing white stars on bursting or an explosive sound signal.

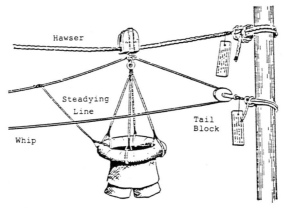

Fig 25.1 Method of securing breeches buoy

26 Inflatable liferafts and personal survival at sea

The inflatable raft is now widely used on board all types of ships as a life-saving facility. They are provided on most modern fishing vessels. Induction courses for British pre-sea trainees include a short course on their use. These courses cover the action to take prior to abandonment of a ship, abandonment, and basic survival technique.

Every crew member must find out where the liferafts are stowed and his own liferaft station. Liferafts are usually stowed and kept in fibreglass containers or valises. The fibreglass container is secured by means of strapping with quick release slip of the Senhouse type. The valise is usually stowed in an upper deck recess.

In the event that a fishing vessel has to be abandoned, hopefully after the distress message has been sent as described in *Chapter 24*, the order to abandon can only be given by the skipper by word of mouth, preceded by a continuous ringing of the alarm bells and/or the sounding of the ship's whistle.

Before abandoning ship, a fisherman should insofar as circumstances allow, put on as much warm clothing as possible (woollen clothing is best) and wear a lifejacket.

NB: Without a lifejacket even good swimmers will have difficulty in staying afloat in cold water because of the disabling effects of cold, shock and cramp. A lifejacket will keep you afloat without effort or swimming no matter how much clothing is worn. If unconscious a lifejacket will keep your mouth clear of the water.

If circumstances permit, blankets should also be taken. The emergency portable radio transmitter and any portable EPIRB should be taken to the liferaft. These are important aids in search and rescue.

To launch an inflatable raft, the painter, which runs out from the container or valise must be properly secured to the ship. The raft is launched over the side and a hard pull on the painter will activate the CO_2 release mechanism.

Inflationary pressure breaks open the raft from inside the container in which it is stowed; the buoyancy tubes which run around the raft are inflated as are the arch tubes which support the canopy.

If the raft inflates upside down, which is unlikely, it can be inverted. Two straps are situated on the bottom of the raft along with a CO_2 cylinder. If someone stands on the cylinder, holding the straps one in each hand and throws his weight backwards, the raft should come to rest the right way up. If the raft is manoeuvred so that the topside or tube arch is facing the wind, the job will be much easier to do. The raft will of course fall on the man but because the bottom is soft and pliable he should not come to any harm.

When the raft is first inflated, excess pressure in the buoyancy tubes

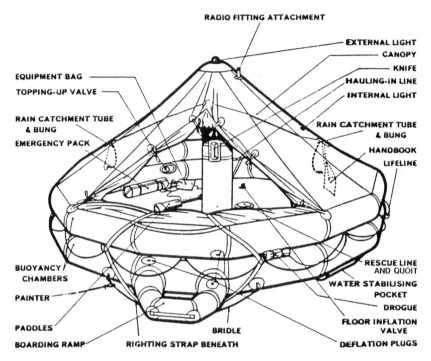

Fig 26.1 Inflatable liferaft. (Courtesy of RFD Ltd)

Note: Positions of equipment may vary

escapes through a pressure release valve. It may sound (to the uninitiated) as if the raft is punctured, but it is not cause for alarm; the screeching noise will stop when the correct pressure is reached.

The raft should then be brought alongside the ship by adjusting the painter to a proper length. It can then be boarded directly from the ship by means of a rope ladder, rope, or scramble net. Entering the water should be avoided. On side trawlers and smaller fishing boats it may be possible to board directly from the deck. One should not jump directly on to the raft except perhaps for the first person since it might be damaged or someone injured. If jumping is the only alternative, the fisherman should jump into the water directly alongside the raft and then board the raft as quickly as possible.

When the raft is full, the painter should be cut with the knife provided and the raft should be moved away from the ship's side. When clear of the ship

The painter has been let out to full length then given a sharp jerk to start inflation

The raft partially inflated.

Wait for full inflation before boarding.

Fig 26.2 Inflating liferaft. (Courtesy Dunlop Marine Safety Ltd)

the drogue or sea anchor attached to the buoyancy tubes near the entrance should be put out so that the drift is reduced. This will assist those conducting a search. The drogue also helps to steady the liferaft. In some models it is so designed as to permit one to align the entrance out of wind and spray in cold climates, and into wind in hot climates. In other models it holds the entrance at right angles to the weather.

If there are other rafts in the vicinity, try to keep together. This can be done by throwing the drogue in their direction and hauling the life-rafts together. Make the liferafts fast to each other on a lazy line to accommodate any sea which might be running. When doing this, look out for any survivors who may be in the water and use the quoit to bring them in. Rafts in company are easier to detect visually or by radar. They also have more survival aids to share.

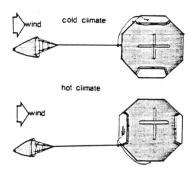

Fig 26.3 Drogue arrangement (Courtesy Dunlop Marine Safety Ltd)

Associated liferaft equipment

The following items of equipment are normally fitted to all liferafts.
Lifeline. Fitted outside and inside the buoyancy chambers.
Righting strap. Provides the means of righting the raft in the unlikely event of it inflating upside down.

Hauling-in lines. At the entrance to the raft as boarding aids, webbing in the form of rungs, or hand holds.

Boarding ramps or ladder. At entrance.

Sea light. Mounted on top of the raft to assist location in darkness. Internal light fitted to arch tube.

Valves. Non return pressure relief valves fitted to each chamber and floor for topping up with pump.

Radio aerial. A fabric sleeve type tube being assembled to the canopy fitted to mount a radio aerial.

Rescue line and quoit. Stowed at entrance attached to a lifeline loop. For use as a heaving line to be thrown to survivors in the water.

Drogue and line. A sea anchor in the form of a drogue and line, folded and stowed adjacent to entrance, outside on a loop patch on the buoyancy chamber. The free end of the line is attached to a loop patch situated at $90°$ to the entrances.

Paddles and sponges. Tied to an anchorage patch on the floor.

Floating sheath knife. Fitted with a buoyant handle mounted on the arch tube or canopy strut.

Liferaft instruction leaflet. Headed 'Immediate Action' on booklet secured and folded inside the canopy of the raft adjacent to the water catchment.

The equipment to be provided in liferafts is laid down in International (SOLAS) Regulations.

The types of packs required are determined by National Regulations. In general these comprise a combination of emergency pack and equipment bag. Their contents are normally marked on the outside. A typical combination would be:-

Emergency pack or drum. Containing: Safety tin opener, first aid kit, drinking vessel, heliograph and whistle, parachute distress signals, hand held signals, fishing kit, ration packs and water, sea-sickness tablets, Immediate Action instructions, SAM suits (where supplied), and baler.

Equipment bag. Containing: Leak stoppers, torch and batteries, repair kit, baler, spare sea anchor, topping up pump, rubber plugs, Immediate Action leaflet and rescue signal table.

Immediate action

First and foremost, protection should be against the dangers of the environment. *Protection* has a higher priority than indicating *location*, and since it is possible to survive many days without *water* and weeks without *food*, both protection and location have higher priorities than food and water. *Do not attempt to sail away from the area of the sunk ship.* Search for survivors will commence at the last known position of the ship.

When everyone is inside close the entrance but leave a small opening for a lookout.

Issue everyone with sea-sickness tablets and take them as early as possible. Most people, including 'hardened' sailors, suffer from sea sickness in survival craft. This results in loss of body fluid and incapacitation.

If any man has been in the water try to dry him and keep him warm. On some vessels there are insulated bags known as SAM suits, which can be pulled on by a man and drawn to around the neck in order to preserve body heat. Use one of these if a man has been in the water and is suffering from exposure (hypothermia).

Keep the bottom of the raft dry by using baler and sponge. *Do not issue any water ration during the first 24 hours in the raft to normally healthy men.* Thereafter issue one pint per day but not a whole pint at once. It should be given out in three or four equal issues at fixed intervals during the day. *Do not drink salt water.* Food rations should be issued in a similar manner.

Being actively employed is good for morale. Watches and duties should be allocated as soon as possible. The raft not only keeps survivors afloat, it protects and preserves life. Because the canopy is double skinned, the buoyancy chambers and the inflated floor surround the survivors with a layer of gas and still air. Keep it this way by having regular inspections and effect repairs according to the instructions in the booklet. Examine the raft for flabbiness and correct by using the topping up pump. Set lookouts and attend to the water catchment bag.

Fig 26.4 Inside liferaft. Maintain the liferaft. Inflate the floor for insulation against the cold, bale out any water and check for damage or leaks. Ventilate the liferaft by maintaining a small opening.

If the position of sinking has been reported, the chances of being rescued quickly will be good. If uncertain as to whether the position has been reported, assume that it has not and initiate distress signals with portable radio or EPIRB when carried. (*Use radio on silence periods if possible*).

In any case of doubt be prepared for a waiting period of several days.

If ships or aircraft are searching for liferafts their attention may be drawn by the use of the pyrotechnics provided in the emergency pack. However, these parachute rocket signals and hand flares are in limited supply and should be used *sensibly*. *Never* ignite them in the hope that they may happen to be seen when the presence of ships or aircraft is not apparent. Wait until an approaching ship is as close as it is thought likely to get. In the case of an aircraft, wait until it is actually in sight and heading approximately in your direction. The heliograph by day and the signalling torch by night will be useful supplements. The empty water tins if hung high in the raft on the canopy will help to make a better target for radar searches and thus improve the detection range. If sighted by a searching aircraft, it may be some time before rescue is at hand but your location will be known.

If a liferaft and survivors make it to shore, the raft should also be brought ashore if possible. The conditions ashore may be such that the raft will still need to be used as a shelter, *eg* in uninhabited, remote areas with extremely cold conditions when immediate aid is unlikely. The raft should be lashed down in a sheltered place.

27 Fire prevention and fire fighting

Fire prevention

Serious fires at sea are best prevented by good ship husbandry, by attention to the correct maintenance of fire fighting appliances and regular fire drills. The crew should be fully aware of their fire stations as set out on the Emergency Station Card which must be displayed on all vessels of 24.4 metres length and over. Every fisherman should ensure that he is familiar with the operation and stowage of the fire-fighting equipment.

The fire alarm signal: continuous sounding of alarm bells and/or ship's whistle.

Accommodation

In fishing vessels bad personal habits, such as smoking when turned in and the careless disposal of cigarette ends, have often caused fires resulting in injuries and loss of life. Everyone should be concerned with keeping the living quarters clean and tidy to reduce the risk of fire.

The following is a list of some precautions which should be taken to avoid fires and fire spread.

DO NOT use cardboard boxes as wastebins, lidded metal boxes are the answer.

DO NOT tie washing lines to electric cables, fire detector or sprinkler heads.

DO NOT use electric light bulbs of more than 60 watts in bunk lights. (Strip lights are safest). Use only correct wattage bulbs throughout the accommodation.

DO NOT rig unauthorised lighting on wandering leads.

Galleys on board ships always present a fire hazard. The cook and his assistant must be fully conversant with the fuel controls and valve settings on oil fired cooking stoves. Other crew members who are likely to use the galley should be instructed in the correct operation of controls and valves.

Pans containing hot cooking fats and oils must be watched carefully.

Engine room

One of the most common causes of engine room fire is an oil leak or spillage. Oil, especially under pressure, leaking on to a hot exhaust, electric motor or generator can ignite and spread very rapidly.

Every precaution should be taken to avoid overflows and spillages when transferring oil or filling tanks.

Oil leaks and spillages should *not* be tolerated. Leaks should be rectified and spillages mopped up immediately whenever they occur. Oily rags, waste *etc* should be disposed of *at once*; drip trays emptied regularly; tank tops and bilges should be kept clean and oil free.

Working spaces

Working spaces should be kept tidy. Rubbish, rags, rope yarns, *etc*, should *not* be allowed to accumulate and become a fire hazard. Fish meal holds should be considered as fire risk areas due to the susceptibility of fish meal to spontaneous combustion.

Fire-fighting equipment

All fire-fighting equipment should be checked for proper stowage and condition by the Mate and Chief Engineer at the start of each trip and at each fire drill. Any faults should be rectified or reported as appropriate. Ventilator flaps and emergency fuel oil 'shut-off' valves should be checked at the same time. Compressed air breathing apparatus sets and spare cylinders, fire detection and sprinkler systems should be tested *every week*. The emergency fire pump should be tested by being operated at every fire drill. All such tests should be recorded.

Fire drills

National Administrations are responsible for making regulations on fire-fighting equipment, training and drills which meet International regulations. In the United Kingdom the Fishing Vessels (Safety Provisions) Rules, 1975 set out the number, type and positioning of fire fighting equipment to be provided for fishing vessels of varying lengths. They also set out regulations on fire drills for such classes of vessels.

Under these regulations it is mandatory that a fire drill be carried out, in vessels of over 24.4 metres, at the beginning of each voyage and at intervals of not more than 14 days thereafter. Should more than 25% of the crew be replaced in any port a muster must be held within 48 hours of leaving that port to ensure that the crew understand and are drilled in their assigned duties. The opportunity should be taken at each drill to give instruction to the crew on the operation and use of fire equipment.

In vessels under 24.4 metres the crew must be made familiar with all fire and life saving appliances and trained in their use at intervals of *not more than one month*.

When drills are carried out, the fact must be entered in the log book. If for any reason fire or liferaft drills are not carried out, this fact should also be recorded, stating the reason.

Where appropriate, Emergency Station Cards should be made out and displayed before sailing.

More generally, in an exercise drill an outbreak of fire should be assumed to have taken place in a part of the vessel chosen by the skipper. When the alarm is sounded the crew should muster as detailed on the station card.

The fire and emergency fire pumps should be started and the main fire pump charged with water. Hoses should be run out and have water passed through them. Fire extinguishers should be unshipped and one used. Compressed air breathing apparatus and smoke mask should be broken out and worn.

When the drill is completed any extinguisher which was used *must be recharged*; a full cylinder must be fitted to the breathing apparatus set; hoses must be drained and air dried before being coiled from the middle and restowed and controlled jet/spray branches checked, lightly greased, placed in a light plastic bag and restowed.

It is essential that each drill should simulate a real fire.

Fire fighting

In the event of a fire being discovered the alarm must be raised *at once* and the Bridge alerted.

The person discovering the fire, after raising the alarm and sending someone to alert the Bridge, should, if it is safe to do so, tackle the fire using the appropriate extinguisher. A determined and skilful attack in the early stages of a fire will, in most cases, prove successful.

When the general alarm of fire is sounded personnel will muster at assembly points detailed previously (*eg* that specified in the Emergency Station Card). From those points, they will be directed to attack the fire or engage in related duties as necessary. The number and position of assembly or muster points will depend on the size of vessel and the number of crew carried.

If the fire is not successfully attacked by the use of extinguishers, or is beyond the capabilities of extinguishers when discovered, a controlled attack will have to be made by the vessel's fire parties using hoses and nozzles or by the operation of fixed installations (*eg* CO_2 flooding in engine rooms).

At a very *early stage* in any fire an effort must be made to confine and reduce it by closing all adjacent doors, port holes, ventilators and other openings, so that the air flow is reduced.

All fires require *fuel, oxygen and heat*. Remove any one of these elements and a fire will be extinguished. By stopping the air flow the supply of oxygen will be reduced and by the correct use of extinguishers the heat will be reduced and the fire extinguished either by cooling or smothering depending on the agent used.

Recommended colour code and use for fire extinguishers:

Extinguisher	Colour	Use
Water	Red	Solid/dry fires, timber, rags, paper, clothing *only*.
Foam	Cream	Oils and fats (liquid fires).
Carbon dioxide (CO₂)	Black	Oils, fats and electrical equipment.
Dry powder	Blue	Electrical equipment, oils, fats, solids.

Each extinguisher should be positioned to cover the potential fire risk for which it is intended.

When a small fire has been discovered, the alarm raised and an initial attack made, other extinguishers should *at once* be brought to the scene. Adjacent combustible material should be kept wet by using a spray to prevent the fire from spreading and if possible removed, Fire and Emergency pumps started, hoses laid out and charged with water and breathing apparatus broken out and worn.

Do not wait to see if the main fire equipment is required.

Solid/dry fires

These fires occur mostly in accommodation spaces and if caught in their early stages can be easily controlled and extinguished by the use of one extinguisher. However, *never* rely on one extinguisher putting out a fire completely; several extinguishers of the appropriate type should be brought to the fire area. Always make *absolutely sure* a fire is completely extinguished by thoroughly soaking the burning and surrounding material.

If a fire has a good hold, a line or lines of hose will have to be used. Heat and smoke rise – therefore, by crawling along the deck the fire fighters will be in the air stream feeding the fire and will find conditions at deck level cooler. Visibility is better at this level and enables the fire fighter to see the fire and so direct the water jet from either an extinguisher or line of hose into the heart of the fire. (See *Fig 27.2*).

When, because of heat and smoke, it is impossible to get into a cabin or compartment, it should be *sealed and all adjacent bulkheads and decks kept*

Fig 27.1 Use of fire extinguisher.
1 Push safety bar aside.
2 Strike knob to activate and direct
   at fire

thoroughly cooled by water sprays. A watch must be kept on ventilation trunk-
ing, false deckheads and panelling in compartments next to the fire area for
hidden spread of fire.

Fig 27.2 Air supporting combustion drawn in at low level

When it is thought that the fire has gone out in a sealed compartment great
care must be taken when opening up. Heat and fuel (either solid or gaseous)
will still be present in the area – only oxygen and/or additional heat is
required to cause an explosion or flash-over fire.

Always have a charged line of hose ready when opening up.

Make sure all electrical power is off and ventilate carefully and
progressively.

Fires in combustible liquids

Fires involving fuel, diesel and lubricating oils, hydraulic oil, paraffin,
petroleum spirit and cooking fats and oils, unlike fires in solids which build

up relatively slowly, occur almost instantaneously, sometimes with explosive violence, and spread very rapidly.

The reader will note from the extinguisher chart that it is inadvisable to use water on fires involving combustible liquids. Water, if used, must be in the form of a spray and used with expertise. Water in the form of jets must *never* be used as these scatter the burning fuel and increase the size and intensity of the fire.

Fires in combustible liquids usually occur in engine rooms and galleys and it is an important first step, apart from attacking the fire, to *shut off* the fuel supply and *prevent air reaching the fire*. All skylights, deck hatches, doors and ventilation systems servicing these spaces should be closed and shut off.

In the event of the engine room being evacuated the space should be sealed and the CO_2 smothering system where fitted actuated by the use of controls and valves sited outside the space. Re-entry should only be made by men wearing breathing apparatus and equipped with a line of hose. Care must be taken when opening up. Entry should be made at as low a level as possible preferably by the tunnel escape if one is provided. Skylights should be partially opened to allow air to be drawn through the tunnel and upwards towards the skylights thus clearing smoke and fumes and increasing visibility.

Small liquid fires

These may be effectively dealt with by spreading sand over the area using a scoop or shovel, by foam, CO_2 or dry powder.

Foam extinguishers

When used on burning liquids direct the stream of foam over the surface to strike a vertical or near vertical surface at the rear of the fire so that the foam builds up and flows back across the fire to form a thick blanket. Do *not* direct the foam into the burning liquid. (See *Fig 27.3*).

Fire in a pan of fat or oil may be effectively tackled by covering with a damp towel or sack and turning off the source of heat.

CO_2 extinguishers

These should be used in figure eight or sweeping motion across the surface of the fire. If directed straight at the fire the velocity of the gas will cause the fire to flare and spread. It may however be directed straight at the fire in electrical equipment.

Dry powder extinguishers

These may be directed onto the burning liquid as the powder leaves the spreader at a low velocity in the form of a falling cloud. It has the advantage over foam that it extinguishes the fire nearest the fire fighter first but there is no cooling effect and it lays only a thin layer of powder over the surface as against the thick blanket provided by foam.

There is still found on many fishing vessels the pistol type of DP extinguisher. This is operated by cocking a hammer and pulling the trigger. The pistol must be aimed low at the base of the fire to ensure the powder spreads. (*Fig 27.4*).

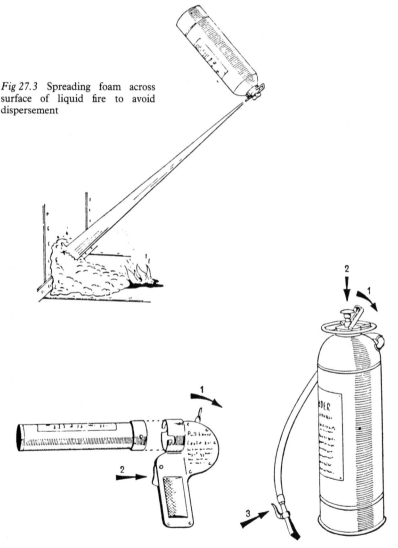

Fig 27.3 Spreading foam across surface of liquid fire to avoid dispersement

Fig 27.4 Small and large powder extinguishers

Electrical fires

All fires of electrical origin should be dealt with by using CO_2 or dry powder extinguishers. Even though it is known that the power is switched off high voltage electrical equipment, such as W/T or radar circuits, water should still not be used. Use *only* CO_2 or dry powder extinguishers.

It is good fire fighting practice in all cases of fire to remove panelling to ensure that a fire is not spreading unseen behind wooden bulkheads and false deckheads.

Automatic fire detection system

There are two types of systems fitted in fishing vessels designed to actuate either when smoke or combustion products are present, or when there is an unusually rapid rise of temperature which is excessively high.

When the detection system is actuated, the fire alarm will sound and a zone light will appear on the detector panel showing from what part of the ship the alarm has originated. The detector panel is usually sited on the ship's bridge. This system also operates the closing of fire doors and smoke doors, and it switches off the ventilator system. Even when alarm systems are fitted, the normal hourly fire and integrity rounds *must* be made.

Sprinkler systems

Some vessels are fitted with these systems in accommodation spaces. The system consists of a pressure tank containing water, control valves, a pump and pipe work terminating with a sprinkler head in each unit of accommodation.

In the sprinkler head is a quartzoid bulb containing a liquid which expands when heated. Also within the bulb is an air bubble, which governs the temperature at which the head operates.

When a fire occurs, the liquid expands, the bubble bursts and a glass valve is released. A jet of water then flows, strikes a plate and forms a spray which helps to control or extinguish the fire. When the head operates, the pressure in the system falls and micro switches operate the alarm signal, activate the sprinkler pump, close fire/smoke doors and stop the ventilator fans.

The sprinkler system should *never* be stopped until the fire is under the control of the fire party and then only on the orders of a ship's officer.

Both of these automatic systems should be tested *weekly* and the tests should be recorded.

Stability and handling of vessel

There is no reason why a shallow skim of water cannot be used on a deck to keep it cool, but in the case of a serious fire where large volumes of water are used, skippers must see that the water is not allowed to accumulate in large quantities, and as far as possible pumps should be used to clear excess water. Free surface water (which is explained in *Chapter 3*) can be *extremely* dangerous. Regardless of whether the compartment containing the slack water is high or low within the ship, at sea or in harbour, large volumes of free surface water may prove to be disastrous. In the past, large ships in dock have been capsized because of the indiscriminate flooding of hold spaces which have caught fire. (See *Fig 27.5*).

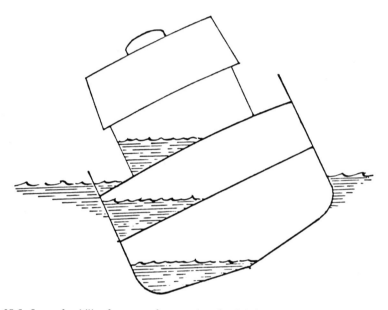

Fig 27.5 Loss of stability from use of water when fire fighting

When at sea with a fire on board skippers will be able to assist the fire fighting party by reducing speed to *dead slow* in order to minimise the relative wind. The skipper may also be able to alter course to put the fire area on the leeside so that smoke, sparks and gases will blow clear of the ship. It will be apparent that a fire under the whaleback would require the wind to be put astern, but any decisions would be made with regard to circumstances, *ie* sea room, adjacent navigational hazards, *etc*.

Skippers are also reminded that a distress or emergency message may be sent depending on the circumstances of the fire and the ability to cope. Other vessels might assist in many ways.

28 Emergencies at sea

An emergency at sea may best be described as a situation which has arisen with little or no warning, putting a ship or persons on board in grave or imminent danger, when only the immediate and correct action taken will be likely to resolve or reduce the gravity of the situation.

Fire and abandonment emergencies have already been dealt with in previous chapters, where it was stressed that the position of all liferafts and fire equipment and its use should be known. An officer who knows his own ship, its capabilities and limitations in bad weather, will avoid many emergencies by taking action in sufficient time to prevent serious damage.

The seaman who knows his ship and is familiar with and understands the pumping arrangements for fuel, water tanks, deck line connections, *etc*, has an immediate advantage in emergencies such as fire, collision, grounding and loss of stability because of an intake of water. If the officer knows the amount and type of gear in the ship, such as blocks, tackles, wires, ropes, sand, cement, boards, tarpaulins, *etc*, then no time will be lost when an urgent situation arises which has to be dealt with quickly.

Explosives

A rare but extreme hazard to trawlermen is mines, torpedoes, depth charges, bombs or other explosive missiles which are sometimes picked up in trawls. The following guidance is given for dealing with them:

— A suspected explosive weapon *should not be landed on deck* if it has been observed before the codend has been opened. In side trawlers, the trawl should be lowered back into the sea and where possible towed clear of regular fishing grounds before cutting away the net as necessary. The position and depth of water where the explosive weapon was cut away, should be passed to the Naval Authorities *via* the Coastguard, ship's agent or owners, or fishery officer, and, if possible, marked by a dan buoy.

— In the event of the weapon not being detected until the contents of the trawl have been discharged on deck, or in the case of a stern trawler, when the codend has been hauled up the ramp, the skipper must decide whether to rid his ship of the weapon by passing it over the side or to make for the nearest port, informing the Naval Authority by radio without delay. His decision will depend on circumstances, but he should be guided by the following points:

— Great care should be taken to avoid bumping.
— If retained on board it should be stowed on deck, away from heat and vibration, *firmly chocked and lashed* to prevent movement.
— It should be kept covered up and damped down properly by wet sacking and the use of a spray nozzle hose. (This is important because any explosive which may have become exposed to the atmosphere *is liable to become very sensitive to shock if allowed to dry out*).
— The weapon should be kept onboard for as short a time as possible.
— If within two or three hours steaming of the coastline of a country possessing explosive disposal resources (*eg* UK) the safest measure will generally be to run towards the nearest port and lie a safe distance off shore to await the arrival of Naval Explosive Disposal Unit. *Under no circumstances should the vessel bring the mine or weapon into harbour.*
— *No* attempt should be made to clean the weapon for identification purposes, open it or tamper with it in any way.
— A ship with an explosive on board, or in her gear, should warn other ships in the vicinity, giving her position and, if applicable intended position of jettisoning.

Drifting mines may occasionally be sighted. They should be reported *immediately* to the Naval Authorities *via* the Coastguard and coast station giving the time of sighting and an accurate position, so that an appropriate warning to shipping may be broadcast.

Man overboard

Much has been written in seamanship books on this subject and how best to turn a ship round and return to the position where the man may be expected. If we take a realistic view of what happens when a man falls overboard, it will be seen that an officer of the watch has very little time in which to take immediate action to avoid the man being drawn into the propeller race.

A trawler steaming at 12 knots will pass a man overboard at a speed of 21.6 feet (6.6 metres) per second. If the man has fallen overboard, let us say from amidships about 100 feet (30.5 metres) from the stern of a trawler at a speed of 13 knots, then he will have been passed in less than five seconds. If the vessel is trawling with a speed of about five knots, he will then have been passed in about 12 seconds.

If the officer of the watch actually sees the man fall overboard, the chances of his being able to stop the propeller in a few seconds are remote. If practicable the wheel should be put hard over *towards the side where the man has fallen* so that the stern may be thrust away from him. Experience shows, however, that a man overboard will generally float clear of the ship, even if no action is taken, by reason of the bow wave and displacement effect. It is important that the man's position be marked and that he be picked up as soon as possible. Loss of body heat is as great a danger as drowning. It is *vitally important* that the bridge lifebuoys be released as soon as possible. The man in the water may be able to reach the lifebuoy for support, but just as important is the fact that the light and smoke signal attached to the float will act as a marker to which the ship may return.

Immediate action

— As soon as the message or call 'Man overboard' is received, release the bridge lifebuoy, with the self-igniting light/flare.
— If practicable put the wheel hard over towards the side from which the man has fallen.
— Set lookouts to keep the man and lifebuoy in sight at all times. Call the Skipper.
— Sound the emergency signal and/or the 'Man overboard' signal.
— Turn the ship round (see remarks below).
— Have the boat or liferaft made ready.
— Inform any ships in the vicinity of the incident.

The best action to take if the man and/or the lifebuoy is in sight is to let the vessel come right round on full rudder, reducing speed towards the end of the turning circle so that the man in the water is brought to leeward for the boat or liferaft to pick up. If the lifebuoy was not released as soon as the man fell overboard, skippers should steam slowly back on a reverse course from the lifebuoy's position. The self-igniting light will burn for a period of not less than 40 minutes, which should give a good departure position from which to search.

Should it become necessary to steam back down the track to search for a man who has been reported as missing and there is no knowledge of when he fell overboard, the following method of reversing course may be used. Put the wheel hard over to port or starboard until the ship's head comes round to but not past 60° from the original course. Now put the wheel hard over in the opposite direction until the ship's head has swung on to a reciprocal course to the original. It will be found that most single screw ships will now be on or close to the original track.

Grounding and refloating of vessels

No two strandings of vessels are likely to be identical and it is not possible, therefore, to produce a set of rules on how to refloat a stranded vessel. The advice given below covers some general principles but the success (or otherwise) of any particular salvage operation will depend upon the degree of planning and skill used in its execution.

Serious damage and possible loss of a vessel may result in attempting to refloat a grounded vessel unless the operation is conducted with proper forethought. The following advice is offered to skippers who may require assistance or who may be asked to assist another vessel.

Action after grounding

When a vessel goes aground, the engine should be stopped *immediately* and should not be moved again until the skipper is satisfied that:

— His vessel is in a fit condition to be refloated and not making water which cannot be controlled.
— Refloating his vessel will not cause further damage to the propeller, rudder or hull.

The skipper's first actions on grounding should be to:

— Sound the general alarm.
— Make ready liferafts and lifeboats if carried.
— Transmit the appropriate distress signal.
— Send the mate and chief engineer to check for damage.
— Alert his owners and the insurance company.

Refloating

Before attempting to refloat, the skipper should take into consideration the nature of the bottom and the state of the tide. If the vessel is lying on a rocky bottom she may be holed and by working the propeller or taking a tow from another vessel the damage may be aggravated.

If the vessel is lying on sand, shingle or mud, it may be possible to refloat immediately. In such circumstances the propeller should be worked astern for short periods only and the engineers should report *at once* if sand or silt is being drawn into the engine intakes. On a falling tide, unless refloating is almost immediate, it will be necessary to wait for the following tide. If the vessel cannot be refloated immediately, soundings should be taken all round her with forward and aft draughts noted.

Jettisoning of gear

Skippers should consider ways of lightening the vessel as a way of refloating her. Jettisoning of ice, catch and fresh water may make an appreciable difference to the draught and jettisoning of fishing gear and anchors, cables, *etc* may also help, particularly if the vessel is aground forward. When jettisoning ship's gear, it should be buoyed for subsequent recovery.

Use of assisting vessel

When seeking the assistance of another vessel, skippers should give careful thought to the procedure to be adopted before making any attempt to refloat. Generally speaking, a vessel aground will refloat most easily if she is taken off in the reverse direction to which she has gone on. This means towing her off astern. Pulling her round by the head should *not* be attempted as this may cause serious underwater damage.

Skippers of vessels rendering assistance must exercise great care when approaching a vessel which has run aground in order to avoid damage to their own vessel. If they are in any doubt about the depth of water in the vicinity of the stranded vessel they should anchor in a safe position and pass the tow by boat, rocket line or other means.

Use of anchors when aground (See also Chapter 29)

It is possible, in good holding ground, to get a better purchase from the use of ground tackle and the winch than from a tug or another trawler. Skippers should, therefore, consider the use of their bower anchors and warps for this purpose if there is a suitable vessel available to lay them out. The anchors should be laid out in a narrow V in the direction in which it is desired to refloat and warps shackled on. Anchors must be buoyed before being laid. The lead for the warp should be kept as low as practicable in the stranded vessel to avoid forcing the stern lower into the water.

If it is not possible for the stranded vessel to get her own anchors away, it may be advantageous for the skipper of an assisting vessel to put his own anchors down having shackled them to the warps of the stranded vessel. The assisting vessel can then tow independently while the skipper of the stranded vessel attempts to heave off.

If circumstances allow, the assisting vessel may be able to drop her anchor in an appropriate line with the stranded vessel, paying out several lengths of cable before connecting up to the grounded vessel. By towing on the warp and heaving on the anchor simultaneously a strong and continuous bollard pull will be exerted on the vessel aground. Experience shows that by keeping the towing warp tight throughout the operation, immediate advantage is taken of any tidal range or surge from ground swell which may take place.

If a skipper is in any doubt as to the best course to be followed after grounding, he should seek professional advice either directly, through his owners, or the nearest agent. In certain circumstances it may be better to wait for professional salvors to arrive than to make a hasty and ill-judged attempt to get off quickly. On such occasions, the skipper should ensure that all openings are made *watertight* and if there is a danger of the vessel working on the bottom, it may be necessary to take ballast on board either in the tanks or by flooding compartments to keep the vessel *firmly* on the bottom.

Beaching a damaged ship

If a ship is so badly damaged that the ingress of water cannot be brought under control, it may be possible to beach her in time to save the ship. The optimum conditions for beaching would be, at or about high water, a sheltered bay with a soft and shelving bottom.

In tidal waters and if there is any choice, it is better to ground the vessel on a falling tide, rather than to force her on to the beach. However, there are many factors which may govern a beaching situation, the main one being how quickly the ship is taking water and the rate at which she is settling in the water.

In an extreme case, such as severe damage following a collision, with a fast intake of water into a large compartment (such as the fish or engine room), the vessel would have to be beached regardless of tidal conditions.

In less urgent cases where time allows, circumstances such as tidal range or whether it is necessary for the damaged area to be uncovered at low water will have to be considered. Vessels should *not*, if possible, be beached too high up on a spring tide. This may make refloating unnecessarily difficult.

Vessel immobilised at sea

A skipper of a disabled vessel which requires towing to a place of safety should:

— Inform his owners and the nearest agent of his insurance company.
— Wherever possible arrange to be towed by a vessel which is insured with the same company as that of his own vessel or with which local port agreements operate.
— Under circumstances where there is no such vessel available and the need for assistance is immediate, skippers should agree terms with the salvage vessel under Lloyd's Standard Form of Salvage Agreement, commonly known as Lloyd's Open Form.

Approach to a vessel requiring a tow

In the open sea, the assisting vessel should *not* attempt to go alongside the ship which is to be towed. A line should be passed either by floating pellets downwind towards the disabled vessel, by firing a rocket line or by towing a line from leeward around the stern of the disabled vessel.

If the disabled vessel is at anchor, going alongside in fine weather in a single screw ship is not a similar operation to that of going alongside a quay, for two reasons. A quay or jetty is a fixed object and cannot move. A skipper going alongside a jetty may make allowances for wind and tide as they affect his ship. A vessel at anchor in a moderate wind will pivot round her stem, to a degree varying with the strength of the wind, tide and swell. (See *Fig 13.7 page 128*).

Even in calm weather, an anchored vessel will pivot if a fine approach is made by the vessel which is to go alongside. The water pressure or displacement effect from the bows of the assisting ship will push the stern of the anchored vessel away and her bows will swing across or towards the line of approach. This interaction will take place even if the disabled ship is being steered, is at anchor, or is using a sea anchor. Therefore the passing of a line by any means is a safer method of making a connection than by going alongside.

Anchor cable use when towed

If a trawler becomes immobilised to the extent that she needs to be towed to port, the anchor cable should be used for towing.

If there is no immediate danger, the towed vessel should bring the anchor and cable under the flare of the bow and onto the foredeck in order to unship the anchor. This will leave the unstudded or open link and the joining shackle free to connect to the eye of the towing vessel's warp.

If time or conditions do not permit the anchor to be unshipped then a successful tow can still be accomplished with the anchor still connected to the cable. In this event *neither* the crown shackle on the anchor shank, *nor* the cable joining shackle, must be used for the connection. If this is done, a sideways pull will be made on whichever shackle is used and *they will distort under the stress*, loosen the pin, the anchor will be lost and the tow will part. Connection of the towing vessel's warps *must* be made by a suitable shackle to the first unstudded or open link. The shackle selected must be of a suitable size and strength and should be properly seized when shackled up. (See *Fig 28.1*)

If it is not possible to shackle to the first open link, then the shackle should be passed around the cable and secured to its own standing part. A trawl door bracket shackle would probably be suitable. Smaller shackles should *not* be used as they will readily part under stress and the tow will be lost.

When the connection has been made, the disabled vessel should pay out two or three lengths of cable, screw up the windlass and put a preventer on the cable. This preventer should be of wire rope and equal in strength to that of the towing warp. A good method of doing this is to arrange the length of cable paid out so that (let us assume) the third cable shackle is on deck near to the inboard end of the hawse-pipe. Set up a threefold wire purchase at the outer link of the third shackle, with the standing part of the purchase on the bitts or bollards behind the windlass. The hauling part of the purchase may then be passed *via* the bitts to the winch and set tight before being stopped off and secured. The brake, compressor or other holding device may also be used but the preventer should bear the weight of the tow.

Connecting and getting the tow under way

When proceeding towards a disabled vessel, the assisting vessel should make

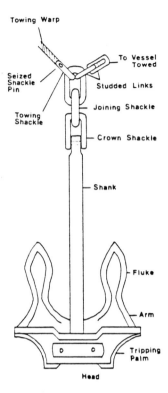

Fig 28.1 Connecting a tow when anchor and cable are to be used

ready all the gear necessary for towing. Rocket line throwing apparatus should be prepared in a clear area free from obstructions. Messenger lines should be flaked out so that they may be run away freely, with a few lengths of warps similarly disposed. Skippers are aware that the modern polypropolene ropes are buoyant so that if floating pellets are attached a quick and efficient connection can be made by floating the line downwind either from the assisting vessel or the disabled vessel.

Tankers – use of rockets

In the event of a trawler having to assist a tanker in a distress situation, it may be found necessary to pass a line between the two ships in order to make up a tow or in order to haul boats or rafts from one vessel to the other.

It may be *dangerous* to establish communication by means of a rocket-throwing apparatus with an oil tanker should that vessel be carrying petroleum spirit or other inflammables. The assisting vessel should lie to windward of the tanker and *ascertain whether it is safe to fire a rocket in her direction before doing so.*

When a vessel in distress is carrying petroleum spirit or other inflammable liquids and is leaking, the following signals should be exhibited to show that it is dangerous to fire a line carrying rocket because of fire risk.

By day: Flag B of the International Code of Signals
By night: A red light hoisted at the mast head

When visibility is bad the above signals may be supplemented by the use of the following International Code Signal made in sound: GU (– – · · · –) 'It is not safe to fire a rocket'.

On coming up to the disabled vessel, time will probably be saved in the long run if the skipper of the towing ship makes a careful and deliberate assessment of the attitude and drift both of the disabled vessel and of his own when stopped. It is therefore a good practice to circle a disabled vessel at close range before attempting to pass a line. If uncertain as to how his own ship will lie and drift, he should stop her clear of the disabled ship and note her behaviour before attempting to pass a line. If it is obvious that the rates of drift are going to be very different, *eg* a side trawler intending to tow a deeper laden freezer trawler, then it is *essential* that the tow is passed as quickly as possible and that everything is prepared before passing the line, as described above.

It is stressed, however, that the state of the weather, condition of the disabled vessel, and above all the need to connect up quickly because of the situation which a disabled vessel may be in, *eg* the proximity of a lee shore, must be taken into account when the skipper is planning the operation.

When the messenger line has been passed between the two vessels, the towing warp should be hauled across by the disabled vessel. The assisting ship should bend the messenger on to her towing warp above and clear of the eye: this will enable those on the disabled ship to make a quick connection to the anchor cable. When passing the messenger line and warp, the assisting vessel should try to maintain a steady heading and distance off, so as to avoid making the transfer unnecessarily difficult. When the warp has been connected to the towed vessel's anchor cable, about three lengths of cable should be paid out and the cable secured, as previously described.

The towing vessel at this stage should exercise great care and patience on going ahead, by intermittent bursts of ahead movement at slow speeds while paying out the towing warp as necessary until about 10 to 12 lengths (250–300 fathoms) of warp have been veered. The greatest stresses on the gear will be made when the inertia of the towed vessel is overcome. The towing vessel will be forewarned of the stress to come by watching the towing warp which will slowly alter its angle between the towing block and the sea, possibly from the vertical upwards towards the horizontal. Engine should be stopped, and, if there is only a little way on the vessel, the catenary (or bight of the towing warp) will become shallow and might even come out of the water. If it appears that the towing warp is to become horizontal, be prepared to pay out more warp from the towing winch so as to avoid a heavy strain on the gear. The length and weight of the warp and anchor cable should overcome the inertia of the towed vessel, the catenary should fall below the surface of the sea, and, if the engine is put ahead again at minimum revolutions and gradually increased, the bight of the towing warp should remain immersed at all times. An aid to overcoming the towed vessel's inertia is for the towing vessel to begin the tow heading down wind at an angle of about 45° to that of the direction in which the towed trawler is laid. Once the tow is moving, speed should be gradually increased until the maximum towing speed is attained.

After the tow is under way, any alteration of course should be made gradually, a few degrees at a time, until the required heading is reached. If it is possible, the vessel being towed should steer, and the propeller should be allowed to trail by disconnecting the shaft between engine and thrust block. The fixed propeller offers high resistance underwater and reduces the speed of the two considerably as well as adding stress to the towing gear.

During a long tow in good weather there are no particular problems as long as the towed vessel rides comfortably astern and is not subject to yawing. The length and weight of the towing gear is all-important and the bight should remain well immersed at all times. If the bight of the tow rope appears in good weather, then it is not long enough and the warp should be veered, if there is a sufficient depth of water. If the weather deteriorates and the catenary shows a tendency to appear then speed should be reduced and/or the tow lengthened.

During the tow, the condition of the gear should be kept under constant observation in both ships for any signs of weakness, so that appropriate action can be taken if necessary. The towing warp should be parcelled by hides or sacking where nipped, and freshened every few hours by veering with an appropriate reduction of speed.

Towing a ship stern first

In the case of a disabled vessel being damaged forward and remaining afloat only because of her collision bulkhead, it may be thought wiser to tow her stern first, in order to avoid further weakening forward.

If the disabled vessel is down by the head, the actual towing of the vessel may not necessarily be more difficult, but the connection may be more complicated from the stern of a disabled side trawler. If possible the disabled vessel should connect to the towing warp by means of a bridle consisting of two lengths of cable or warp of equal length, one from each quarter. The rudder should be set in the midships position and secured.

If the disabled vessel is deeper aft than forward, the use of a second vessel connected astern of the tow may be necessary in order to maintain a course and prevent excessive yawing.

Hull damage

It is vital to reduce the inflow of water from hull damage whenever practicable, even if the means used do not stop it completely: it must, when possible, at least be reduced to an amount the pumps can cope with. For example a hole 6 inches in diameter 6 feet below the surface would allow an intake of sea water of about 242 tons per hour. Speed is therefore imperative in controlling flooding. Also the deeper a vessel sinks in the water the greater will be the water pressure on the leak. In all cases of leakage, engine room and portable pumps should be brought into use immediately; and if necessary baling by buckets hand to hand should be organised.

It is often very difficult to get at leaks in the hold of a fishing vessel from the inside because of the fish stowed there; in engine rooms also because of the location of fuel tanks.

Small holes (such as a rivet hole) may be plugged by knocking a tapered broom handle into the hole. When this has been done a cement box should be fitted around the hole and left to harden. If a small hole has been caused by corrosion, great care should be taken when plugging, a split or fracture might result if the plug is hammered in too heavily. A more permanent repair may be made as follows:

To act as a float use a cane or stick, as long as is convenient, of a diameter allowing it to pass freely through the hole and attach to it the end of a ball of twine. Push the stick through the hole and pay out the twine until the float reaches the surface of the water. Pick up the float and send down a length of marline or light line by hauling in on the twine. Have a threaded bolt with a good head on it, a large washer and a gasket made of rubber or other suitable jointing fitted on to the shank of the bolt. Make the head of the bolt fast and by using marline hitches on the bolt shank pass it down to the hole where it may be pulled inboard. Have a suitably threaded nut, washer and gasket ready to fit once the bolt is inboard and the marline is clear. Put on plenty of red lead and tallow and screw up tight. It should be unnecessary to make a cement box on this type of repair if the gaskets and washers cover the hole.

Cracks which are not too wide may be covered from the outside of the ship by cowhides, blankets or tarpaulins. Firmly secure the material to be used between two lengths of lines. By passing the bights over the bow and moving aft, the cowhide, blankets or tarpaulin may be manoeuvred until over the crack in the hull. By spreading the lines carefully and setting them tight; the crack will be covered and the water pressure will help to seal the fracture by pushing the material into the hole. Under reasonably dry conditions a cement box can then be made.

Cement boxes may be used after leakages have been reduced as much as possible by use of plugs, wedges, oakum, bedding and tallow. If the leak is stopped completely by these methods, a cement box may be applied immediately. It is always advisable to use the ship's frames when making a cement box. Pound boards may be cut to size so that they fit tightly and wedged between frames above and below the damaged part. If the damage extends across a frame, then two sets of boards above and below the damage can be fitted. Allow a good overlap over the damage and fill with mortar. Cover the mortar by nailing boards over the surface of the mortar to the wooden framework.

When mixing sand and cement use two parts of sand to one of cement. Gradually add water, into which soda has been dissolved, until a stiff but workable mix has been made. All the surfaces to which the mortar is to adhere should be clean and free from grease and running water. If running water from the plugged damage is still present, a channel, tunnel or piping should be laid so that the cement box can be made. When the cement box has set, the channel or piping may then be plugged. Ship's husbands and skippers should always make sure that quick-drying cement is shipped when stores are taken.

Collision damage

In most cases of collision, except that suffered by a bulbous bow, the fracture begins above the waterline and extends downwards in a 'V' shape with localised fractures and indentations. When the damage has taken place in smaller compartments such as a fuel tank, fore peak, after peak, *etc*, it is vitally important to seal off the damaged compartment by closing all openings such as doors, hatches and pipe lines which might cause flooding in another area. Having isolated the compartment, use of a collision mat as described previously may allow the pumping out of the damaged area or at least keep flooding under control. Sound judgment, with quickly taken action, may avoid flooding and subsequent abandonment in cases where damage is adjacent to, or just below the water line. The transfer of fuel and water, *etc*, may be sufficient to list or trim the vessel and so incline the vessel in order that the damaged area clears the water line to permit plugging. The transfer of fuel or water will not, in itself, reduce the reserve buoyancy of the ship, except for the effect of the free surface when the transfer is taking place. The flooding of a side tank from the sea carried out to list the vessel will reduce stability by increasing the draught and lowering the reserve buoyancy. Transfer of fuel and water *must* always be regarded as the safer method of creating a list in a vessel which has been damaged and is already subject to an ingress of water.

Extensive damage to hull

A vessel which has suffered extensive damage to a large compartment may well find that the resources on board are inadequate to effectively deal with the situation. The ship which remains afloat after such damage to a large compartment should make for the nearest port or anchorage either by her own power or by being towed. Do *not* try to reach a more distant port where full repair facilities are available for the following reasons. Because of flooding, the ship's stability will have been reduced considerably by loss of buoyancy due to the extra weight of water on board and free surface effect. The deeper the ship sinks in the water, the greater will become the pressures on the bulkheads within the ship and risk of rupture. If conditions deteriorate, the possibilities of beaching will be more likely if approach is being made towards the nearest port or anchorage. Temporary repairs may be effected at the nearest port which will allow the vessel to proceed to a port with good repair facilities.

Harbour approach

If a successful tow is nearing completion and a harbour or estuary is being approached, the towing vessel should reduce speed in plenty of time and shorten the tow. The ship being towed will not sheer too much on a short towing bridle and will be conducted more safely in narrowing waters where other shipping may be encountered. Correct towing signals *should always be exhibited* by day or night when towing or being towed.

Wherever possible, the towing vessel should hand over the towed vessel to harbour tugs for berthing before entering a harbour or a narrow tidal estuary. If, however, there are no harbour tugs available the assisting vessel may have to go alongside the towed vessel. Great care should be exercised in the mooring together of the two vessels which is preferably delayed until calm water is reached.

Having taken the way off and slipped the tow, the towing vessel should turn round and approach the tow from astern with very little headway in such a manner that when abeam of the tow the distance between the two vessels will be about 40 feet in parallel. Heaving lines should be passed, then breast ropes, so that the vessels may be gently brought flat alongside of each other's fenders. If, when this operation is being carried out, the disabled vessel sheers because of wind or tide, the assisting vessel will, because of her distance off, be able to act appropriately or steam away altogether. In any case the bringing together of two vessels in any sort of sea or swell *should be avoided*, and the operation should be carried out in reasonably calm conditions. Vessels of different lengths, period of pitch and/or roll, because of different stability, will cause damage in spite of fenders when there is any sea or swell running.

Shoring of bulkheads

If resources on board permit, bulkheads may be stiffened by shoring with timber or other means. If for example the bows of a ship are extensively damaged to such an extent that the forepeak chain locker, *etc*, are flooded and the ship is down by the head, it may be considered necessary to stiffen the collision bulkhead which separates the next compartment from the flooded area. This will allow the ship to steam towards the nearest port with the collision bulkhead strengthened against the pressures to which it will be subjected when underway.

By setting up vertical timber pillars between a conveniently placed deckhead thwartship beam and the deck, horizontal shores may be set up against the bulkhead to act as stiffeners. The vertical pillars should be chocked off and braced at the deck and the horizontal stiffeners to the bulkhead should be wedged and set up against planking placed athwart the collision bulkhead. Two or three sets of stiffeners should be set up across the width of the bulkhead in order to strengthen the bulkhead evenly. Any improvised timber buffer work, well set up, chocked and stiffened evenly will strengthen a bulkhead against pressures from outside. Progress to the nearest port should be made at a moderate speed having due regard to the damage.

Part VI – Anchors and cables – cordage and its uses

29 Anchors and cables

Following is a list of terms used in anchor work:

Weigh anchor. To heave in the cable until the anchor is broken out of the ground and clear of the water. The anchor is aweigh when it has broken out of the ground but is not clear until out of the water and it can be seen that there is nothing attached to it. Otherwise it is foul.

Lead or grow. The direction in which the cable leads between the hawse-pipe and the anchor. It is indicated by the anchor officer pointing in the direction in which the cable leads, *ie* ahead, astern, abeam, *etc.* When the cable is underfoot and leading vertically it is said to be 'up and down'. At this stage, when weighing anchor, a slowing down of the windlass usually indicates the breaking out of the anchor from the ground.

Short or long stay. The cable is at short stay when it leads downwards from the hawse-pipe towards the vertical and is at long stay when taut and leading close to the horizontal.

Brought-up or come-to. A ship has brought-up or come-to when the operation of dropping the anchor has been completed, the cable has tightened, the anchor has held, and the ship has come towards the anchor with the cable lying along the bottom and the ship lying comfortably at rest.

Veer or walk back. To pay out the cable by using the windlass in gear and power. The cable cannot thus run out.

Surge. To let the cable run out on the brake and without power.

Snub. To brake sharply when surging.

Clear hawse. When both anchors are out and the respective cables are clear of each other.

Foul hawse.   When both anchors are out and the cables are crossed or turned round each other.

Windrode.   A vessel is said to be windrode when she is lying head to wind regardless of any tide.

Tiderode.   A vessel is tiderode when she is lying head to tide regardless of any wind.

Leetide.   Wind and tide together acting on the anchored ship. Surface water appears unbroken or calm.

Weathertide.   Tide against wind. Surface water broken and choppy.

A-cock-bill.   Or anchor clear to let go, means an anchor 'walked back' so that the shank is clear of the hawse-pipe and ready to let go.

Cocked anchor.   An anchor is said to be cocked when it has been weighed and the shank has entered the hawse-pipe either at an angle (with a part turn in the cable) with one or both of the flukes pointing inboard and on the ship's side. If, by walking back, the anchor does not clear then a wire messenger round the flukes should be used to correct. *Do not* continue to heave in with the flukes pointing inboard.

Anchoring

Before anchoring, choose a suitable position on a large scale chart with a line of approach and tide right ahead if possible. If the tide is on one bow or the other allowance will have to be made on the approach run. Choose the anchoring position bearing in mind the vessel's draught in relation to low water depths and swinging room.

   Note the direction of the wind; see if there are any other ships at anchor and the direction in which they lie to their cables. This will give guidance on how to position the vessel when letting go the anchor. Having chosen a position seek out a conspicuous object or light ahead of the approach line and if possible another mark abeam of the anchoring position.

   With engine room and hands at stations, proceed along the approach line, losing headway as the anchoring position is neared. Be sure that the anchor is ready to let go and when the desired position is reached, either go slow astern or let the tide carry the ship back so that a little sternway has been gathered. Just before the required amount of cable is paid out give a touch ahead in order that the ship is brought up easily by the cable. Do *not* let the cable become bar-tight between the anchor and the hawse-pipe, because of sternway; you may break out the anchor from the ground or part the cable at the hawse-pipe.

Do not as a rule drop the anchor when moving ahead over the ground, except when obliged to do so or when making a pre-arranged running moor. Remember that the cable will stand a tremendous strain when taut and straight, but when nipped in a hawse-pipe or around the stem, it will weaken or break.

There is no firm rule of thumb method on the amount of cable to use when anchoring that will meet all conditions. A popular guide is that the amount of cable to use should be about five times the depth. This is a very rough and ready guide and would only be sufficient for ideal conditions such as: (*1*) Good holding ground; (*2*) Fine weather and little tide; (*3*) Deep waters and heavy cable.

Bad holding ground is usually of rock, stone or very soft mud. Good holding ground is of sand, shingle or soft clay. The most important point to emphasize is that an anchor is *most efficient when subjected to a horizontal pull along the seabed*. Consequently, sufficient cable must be used so that a bight or catenary is made between the ship and the anchor, resulting in a substantial length of cable lying along the sea bed and exerting a horizontal pull on the crown of the anchor. It will be apparent why the heavier cable of wrought iron used on older vessels is better for anchoring than the stronger but lighter forged steel cable now widely used on the more modern ships. All these various factors should be allowed for when deciding just how much cable to use when anchoring, *eg* if anchoring in a river with a muddy bottom (bad holding ground) and a strong tide, the '5 × depth' rule may quite easily become the '8 or 9 × depth' rule.

At anchor

The ship, having been brought up, should now be left to ride to the anchor with the cable secured by the compressor or cable stopper with the brake screwed up as a preventer. If possible the cable should be secured so that the shackle is available inboard from the compressor or stopper. The anchor cable may then be easily buoyed and slipped in case of emergency.

Anchor bearings should be taken, entered in the log book and put on the chart. Convenient bearings abeam should be noted for visual checks against dragging, and a radar bearing and distance should also be taken in the event that visibility deteriorates. Appropriate bridge and engine room watches should be maintained, and officers should be particularly alert when the vessel is about to swing. Apart from the fact that the anchor may be broken out of the ground when swinging rapidly, the engine may be needed, because of other vessels at anchor or because of a lack of swinging room. At night a sure sign that the vessel is dragging will be continuous vibration noises coming from the cable between the outboard end of the hawse-pipe and the windlass. Vibrations can be felt by placing a hand on the chain.

Weighing anchor

With engine ready, the operations of heaving away should begin with the mate in charge forward. A man should be stationed in the chain locker to stow the cable as it comes inboard. Failure to stow the cable properly may result in the chain piling up until it reaches the spurling pipe, thus creating a blockage at the deck head just under the windlass. If the chain locker is too deep for this to happen, the next time the ship rolls the cable will fall over to one side of the locker and may become foul on its own parts. This will cause a blockage and will probably not be discovered until the next occasion on which the anchor is dropped.

If it becomes necessary to veer the cable or if after weighing anchor it is decided to drop the anchor again, then it is the mate's responsibility to make sure that the man stowing the cable *is out of the locker* before any cable is paid out. When heaving in, the mate should indicate to the skipper on the bridge the direction in which the cable leads, so that the propeller and helm can be moved appropriately, allowing the cable to come into the hawse-pipe without being turned around the bow or subjected to undue strain.

Losing an anchor

If for any reason an anchor and cable are lost, the position should be buoyed. If it cannot be buoyed, accurate bearings of the position should be taken to assist in the recovery of the lost anchor. The loss should be reported to the owners and, if within the precincts of a port, to the harbour authority. An *accurate* position is needed to effect recovery.

Mooring

The mooring of a ship becomes necessary when there is a lack of swinging room. If it is assumed that a ship is to anchor where four lengths of cable are necessary but there is a lack of swinging room, she would have to be moored with one anchor leading ahead with four lengths out and the other anchor leading astern with the same length of cable. The ship would turn with the tide from one anchor to the other and pivot in one position.

Standing moor

The vessel should be brought head to the tide and if she has a right-handed propeller, the port anchor be let go at the appropriate time and place. With the ship falling astern and head to the tide, eight lengths of cable should be paid out until she has been brought up. The starboard anchor should then be let go and, with the ship put ahead, it should be paid out to four lengths.

Simultaneously, the port anchor should be hove in until the vessel is riding on the port anchor and there are four lengths out to each anchor.

Running moor

Assuming the same conditions and requirements as before, the running moor is carried out by dropping the starboard anchor first and then paying out eight lengths while going ahead against the tide before dropping the port anchor. The vessel is then allowed to fall back on the tide until she is middled between the two anchors and rides to the upstream or port anchor.

If one had a choice on which moor to carry out, the standing moor should be used in preference to the running moor. The running moor requires that the lee or downstream anchor be dropped first making it necessary to render out astern double the amount of cable finally required while the ship is steamed ahead against the tide. The leeward cable can quite easily be nipped or strained when it is being snubbed out and running astern from the hawse-pipe, and may be damaged or parted if not handled very carefully.

When at anchor on either a standing or running moor, steps should be taken to ensure that the ship swings within the same arc on consecutive tides to avoid crossing the cables. Propeller and helm should be used as necessary when swinging from the riding anchor to the lee anchor.

Open moor

This moor is different from those described previously in that the vessel rides to two anchors lying ahead and angled outward, one from each bow. It is used where it is known that the wind or current will come strongly from one direction and it is necessary to use two anchors.

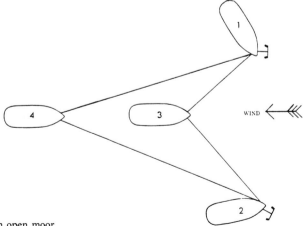

Fig 29.1 An open moor

If we assume that the wind or current comes from the north and the ship is approaching from the west, then the port anchor should be dropped with the vessel headed at a slow speed in a NE-ENE direction, so that an easterly course will be made good in the prevailing wind or current.

With the propeller working easily ahead and the use of helm to maintain an easterly direction, pay out about one third or one half of the required length of cable on the port anchor, and hold on. With little or no headway, bring the ship's head up towards the wind or current and let go the starboard anchor. With the propeller stopped, allow the ship to fall back and pay out the starboard cable until it equals the scope of the port cable, then pay out the required lengths on both, until properly brought up.

It will be seen from *Fig 29.1* that as more cable is paid out on both anchors from position 3, to position 4, the angle between the cables becomes less. The stress on the cables reduces directly with the angle of the cables at the bow so that it can be seen that two anchors, spread sufficiently to avoid fouling, will share the stress better than anchors which are spread too far apart.

Sea anchors

A sea anchor is used to lessen drift to leeward and to keep the vessel out of the trough of the sea when the engines or steering gear are out of action. When such circumstances arise, consideration should be given to anchoring the vessel by her main anchors and not by sea anchors, but if there is no danger of the ship driving ashore and the water is comparatively shallow, it may be possible to keep the vessel out of the trough by unshackling the anchor and paying out the cable until two or three lengths are dragging along the bottom. The drag on the cable should keep the ship's head to sea. This method has been used with good result on numerous occasions. In deeper water, warp may be used to veer sufficient cable to the bottom in the same way.

A string of bobbins through the bow fairlead or from the fore gallows will act as a sea anchor; a trawl door may be used similarly.

Laying out an anchor to assist in refloating

The kedge anchor may be taken away from the ship by boat and will serve to move the ship from one spot to another providing she is afloat and not hard aground. The bower or 'snow' anchor will prove much more efficient in the latter case.

Having decided on which bearing to lay out the big anchor, the kedge anchor should be taken away in a boat, together with plenty of dan wire and a buoy or some pellets. The small anchor should be dropped on the bearing selected for the big anchor, but at a much greater distance from the ship. The buoy or pellets should be made fast to the dan wire in order to mark the position of the anchor and to keep some of the wire from the bottom. The boat should then return to the ship paying out the dan wire, the end of which should be passed back onboard the ship and secured. In returning, the boat must make plenty of allowance for wind and tide because, with a bight of wire on the bottom, it may be difficult to reach the ship. In modern trawlers which do not have suitable boats, shore boats should be used for this work. (See *Fig 29.2*).

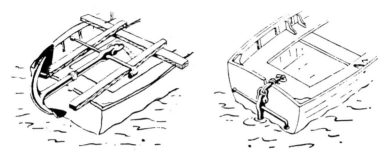

Fig 29.2   Light anchor stowed on the transom and over the stern

From now on the boat can be hauled to and fro on this wire and the use of oars should not be necessary. Thus much time and effort will be saved.

Meanwhile, the big anchor, warps, wire cables, suitable shackles, dan wire, float pellets, pound boards, spars, small line, hangers, seizing wire, hammers, spikes, chisels, saws, nails, wedges, knives, axes and all else should be prepared.

Much depends on the depth of water as to how the anchor should be carried, but when only a single boat is available it will be best, if possible, to sling the anchor under the boat either horizontally or vertically (*Fig 29.3*).

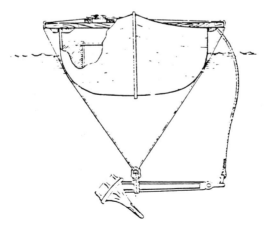

Fig 29.3   Heavy anchor slung beneath a boat

If two boats are available the anchor may be carried between the boats (*Fig 29.4*). Great care is necessary in the slinging of the anchor and preparation of the gear and there must be a wire cable on the anchor with the other end buoyed with dan wire and pellets before the anchor leaves the ship. This makes recovery much easier if the anchor should accidentally be slipped.

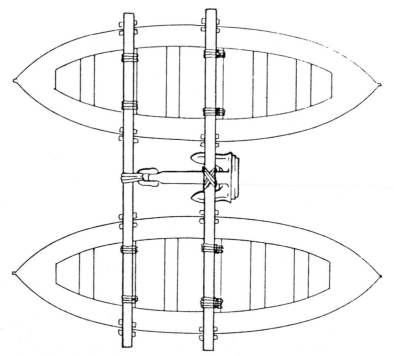

Fig 29.4 Heavy anchor slung between two boats

Some bower anchors are too large to be carried by a small boat.

While the preparations are being made, the boat should be employed in laying out wire towards the spot where the anchor is to be let go.

A certain amount of warp can be coiled in the boat and some can be paid out from the ship as the boat hauls along the dan wire. When the boat has gone as far as she can in this way (and it will not be very far) the coils in her can be paid out, the end being buoyed with dan wire and pellets. There is some danger to personnel in this manoeuvre as the coils in the boat are liable to take charge unless suitably secured and the wire controlled in its run-out by means of a check stopper (*Fig 29.5*).

A trawler commonly carries several wires in lengths of 25 fathoms and upwards to 50 fathoms, such as the wire cables used in fishing. It will be a fairly simple matter to lay out these wires one by one, buoying each length until the next one is brought to it and shackled on.

It may seem laborious to have the boat going to and fro so many times and to shift the dan wire and pellets from length to length but in 200 yards there are only two 50-fathom lengths or four 25-fathom lengths and the job will be done safely and more quickly than if the men have to struggle with the weight of warp involved in the other method. Light loads and quick work is the best rule here.

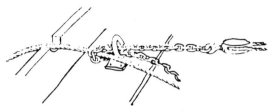

Fig 29.5 Chain check stopper used on a hawser in a boat

Having laid out the wire to a distance of not less than 200 yards, the boat must now take out the big anchor.

First, a strongback must be fitted across the boat to prevent the gunwhales being squeezed in by the weight of the anchor. (See *Fig 29.3*) The pound board may be cut either to fit snugly inside the gunwhale or to lay across the boat and to have a notch cut in each end to accommodate the anchor sling. In this case, wedges or blocks of wood are nailed on to the strongback to fit inside the gunwhale. This latter system is recommended as being easier to fit but the wedges must be very firmly fixed and the strongback lashed securely to the thwarts so that it cannot move.

The anchor will have been lifted horizontally by its point of balance where, if there is no special provision for lifting, a good lashing and two strops will have been placed. One strop is for use with the sling round the boat described later and the second and much longer strop is for taking the whip or tackle hook. The lashing must be prevented from sliding along the shank of the anchor by passing a line from the lashing and across the crown of the anchor. If the water abreast the fore derrick is not deep enough, the anchor must be slung from the after gallows or boat's davit and if the water is not deep enough anywhere alongside, arrangements must be made to lay the anchor on the boat following the same plan as is outlined for the light anchor.

If the anchor is to be slung vertically under the boat the work will be a little easier but the depth of water alongside and all the way to the place of dropping the anchor must be checked.

Before the anchor is swung outboard a wire cable must be shackled to the ring and the other end buoyed with dan wire and pellets. Then, in case of accident, the anchor will not be lost. Also, the strop into which is hooked the whip or tackle to lift the anchor must be capable of being cut unless the hook can be tripped.

As the anchor is lowered to the water line, those in the boat will pass a suitable rope or wire round the boat and through the shorter strop and work it around the strongback and secure it by a suitable bend or by slip and shackle so that the anchor is slung horizontally and centrally under the boat. The tackle may then be disengaged, perhaps by cutting the strop, and the wire cable with its buoyed end passed into the boat, which is then hauled out to the spot where the anchor is to be dropped.

On its way out, the boat will arrive at the pellets marking the end of the connected wire cables already laid out. This end can be hauled up by means of the dan wire on it and can be shackled to the end of the buoyed wire cable in the boat. The cable can be paid out as the boat is hauled further out and then the sling around the boat can be slipped or chopped and the anchor will go to the bottom.

The dan wire and pellets on the last length will serve to mark the spot and make recovery easier.

Steady tension by winch and the anchor will prevent the ship from going further on shore and may pull her off when the tide serves. An anchor so used will stand more pull than two good tugs. If time permits, another anchor can be laid out similarly.

It may be possible to lay out the kedge anchor and to have on it a block through which is rove a dan wire, one end of which can be made fast to the buoyed end of the warp and the other end taken to the winch and used to haul the warp out to the kedge anchor. The boat could then take out the big anchor and one length as before, haul up the warp, shackle up, pay out and then slip the anchor.

Reference has been made to shore boats because the ship's boats may be too small to carry a large anchor in any of the methods suggested. Larger boats, suitable fishing vessels or other craft may be available on hire from near by.

The laying out of an anchor may not always be necessary. The wire cables or warps led out from the ship may be taken to a suitably placed large rock or big tree on the other side of a river, fjord or harbour, or a special holdfast may be constructed on shore.

Chain cable – metric formulae for breaking stresses

The diameter D is expressed in millimetres, the breaking stress in tonnes

Stud link chain		*Breaking stress factor*
Grade 1	(12.5 mm to 120 mm)	$20 \times D^2/600$
Grade 2	(12.5 mm to 120 mm)	$30 \times D^2/600$
Grade 3	(12.5 mm to 120 mm)	$43 \times D^2/600$
Open link chain		
Grade 1	(12.5 mm to 50 mm)	$20 \times D^2/600$
Grade 2	(12.5 mm to 50 mm)	$30 \times D^2/600$

30   Types of cordage and its uses

Fishermen's work involves great use of many types of cordage; steel wire ropes, natural fibre ropes and synthetic ropes and twines.

Steel wire ropes are mainly used for trawl warps, dan leno wires and gilson wires; fibre and synthetic ropes are used for many components of trawls and fishing gear. Many ropes on the trawl are combination ropes being made up of both fibre and wire yarns. Synthetic ropes are also used for rigging trawls and also for towing other vessels. Synthetic twine is principally used for nets. Coir ropes are used where light and springy rope is required.

When selecting the kind and size of rope best suited to a particular use in their vessel's fishing gear, skippers should supplement their own experience by manufacturers' up to date advice on the very wide variety of modern cordage available, notably synthetic, with features well adapted to the many special fishing requirements.

Breaking strains of various sized ropes depend on their basic specification and size, synthetic having substantially greater breaking strains than manila or sisal.

A general guide to breaking strains of ropes in sound condition and allowing a good safety margin can be found from the following formulae:

Metric formulae for breaking stresses of natural and synthetic fibre ropes and steel wire.

The diameter D is expressed in millimetres, the breaking stress in tonnes.

Fibre rope 3-strand hawser laid		*Breaking stress facor*
Grade 1 manila	(7 mm to 144 mm)	$2 \times D^2/300$
High grade manila	(7 mm to 144 mm)	
Polythene	(4 mm to 72 mm)	$3 \times D^2/300$
Polypropylene	(7 mm to 80 mm)	
Polyester (Terylene)	(4 mm to 96 mm)	$4 \times D^2/300$
Polyamide (Nylon)	(4 mm to 96 mm)	$5 \times D^2/300$

Flexible steel wire rope

6×12	(4 mm to 48 mm)	$15 \times D^2/500$
6×24	(8 mm to 56 mm)	$20 \times D^2/500$
6×37	(8 mm to 56 mm)	$21 \times D^2/500$

These formulae afford broad indications. More precise details of particular ropes are given in *Appendix 5*.

As an approximation the safe working load should not exceed one sixth of the breaking strain. This applies also to wire rope used for slings or running gear.

In handling wire ropes it is *most* important to avoid kinks and also to avoid working them over sheaves or barrels of small diameter.

If a rope is spliced its strength is reduced by 1/6th and by as much as a half if knotted.

Artificial fibre ropes

Artificial ropes stretch under strain and it is advisable to stand well clear when they are under heavy loads as they tend to 'fly' rather more than natural fibre ropes when they part under these conditions. They also slip more easily round bollards and winch drums, making it necessary to take extra turns when using these ropes. They also tend to melt when subjected to heavy friction.

They can be spliced like manila or sisal, the end of each strand being whipped before starting the splice, and after the standard four tucks have been taken it is as well to halve the strands and seize the halves together in the manner known as 'shouldering' or 'dogging' to prevent them from drawing under strain.

Each kind of artificial fibre rope has its own advantages over ropes made of natural fibres. Some of them are considerably lighter and some will float and will not absorb water. All those made for work at sea in ships, used with normal care, will give good service.

All ropes are damaged by excessive strain and by chafing and by taking excessive loads with a jerk. When attempting to tow-off a grounded ship, for example, or other difficult tow requiring maximum power, the towing vessel(s) should never lunge into the effort. The two lines must be tautened steadily so as to avoid parting.

It is important to avoid contaminating any rope with oil, both to preserve natural fibres and to prevent synthetic ones from becoming slippery.

31 Tackles

A tackle is a system of blocks through which is rove a rope. It is used to move heavy weights.

The mechanical advantage (MA) of a tackle is estimated by counting the number of parts of the rope at the moving block. It is not always possible to use a tackle to advantage, *ie* to have the hauling part of the fall coming away from the moving block, but the seaman should try to make the block with the greater number of sheaves the moving block.

The greater the MA the less will be the load on the standing block which is calculated by adding the pull required to the weight being moved. Friction is mentioned later and is not taken into account in the following examples.

Single whip

The single whip (*Fig 31.1*) has one standing block and no moving block. (No MA). The load on the standing block is twice the weight to be moved.

Runner

A runner (*Fig 31.2*) consists of a rope rove through a single moving block. (MA = 2).

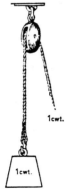

Fig 31.1   Single whip

Fig 31.2   Runner

Double whip

The double whip (*Fig 31.3*) consists of two single blocks with the standing part of the rope made fast to the upper block. (MA = 2).

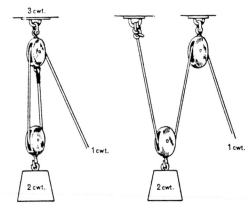

Fig 31.3 Double whips

Luff

The luff (*Fig 31.4*) uses a three inch rope or larger. It consists of a double and a single block with the standing part made fast to the single block.

(The MA = 4 if the tackle is rove to advantage or 3 if rove to disadvantage).

Jigger

The jigger is the same as a luff using a 2 to 2½ in rope.

Handy billy

This is a small tackle like a luff but using a smaller rope than 2 in.

Two-fold purchase

The two-fold purchase (*Fig 31.5*) uses two double blocks (MA = 5 or 4).

Three-fold purchase

The three-fold purchase (*Fig 31.6*) uses two treble blocks. (MA = 7 or 6). The purchase in the illustration is rove to disadvantage and has six parts at the moving block.

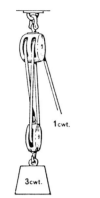

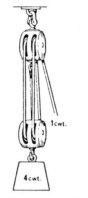

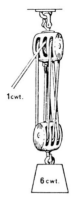

Fig 31.4 Luff   Fig 31.5 Two-fold purchase   Fig 31.6 Three-fold purchase

Runner and tackle

(MA = $2 \times 4 = 8$) (*Fig 31.7*).

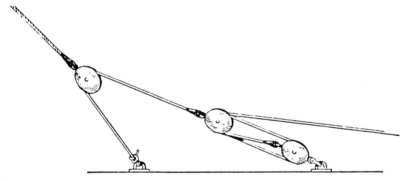

Fig 31.7 Runner and tackle

Luff upon luff

The name describes the combined use of two tackles. The two luffs in the illustration (*Fig 31.8*) are rove to advantage (MA = $4 \times 4 = 16$).

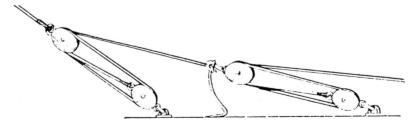

Fig 31.8 Luff on luff

Dutchman's purchase

The Dutchman's purchase (*Fig 31.9*) is a tackle used in reverse to drive a light whip at a fast speed from a slow source of great power. Useful in salvage work.

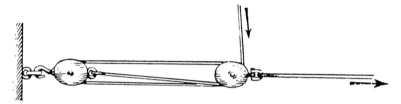

Fig 31.9 Dutchman's purchase

Friction and pull required

In all the assessments of mechanical advantage the effects of friction have been ignored. Friction has to be overcome in the lifting of weights and it is therefore sensible for any tackles or other blocks and sheaves in the ship to be regularly attended and to have the pins of the sheaves well greased at all times.

A rough rule for finding the pull (P) required on the hauling part of a tackle is as follows:

Add the weight to be moved (W) to 1/10 of W for each sheave (S) in the system, *ie* $W + (S/10 \times W)$. Divide the result by the number of parts of fall (F) at the moving block. Thus,

$$P = \frac{W + (S/10 \times W)}{F}$$

In luff upon luff to best advantage, the pull required to move a weight of half a ton is found as follows:

$$P = \frac{1120 + (6/10 \times 1120)}{8} = 224 \text{ lb}$$

By this formula, using one luff only, the sum works as follows:

$$P = \frac{1120 + (3/10 \times 1120)}{4} = 364 \text{ lb}$$

Racking a tackle

If for any reason it is found necessary to move the hauling part of a tackle *ie* from a winch drum to belaying pins or bollards, the tackle may be racked or stopped off by the use of a light rope or seizing cross-whipped around the standing part and a moving part as shown in *Fig 31.10*.

Choking the luff

Another method of stoppering off a tackle temporarily by allowing the hauling part to jam under a moving part of the sheave. The luff is choked by the weight on the hook (*see Fig 31.11*). This method should only be used in an emergency when reasonably light weights are being lifted by hand.

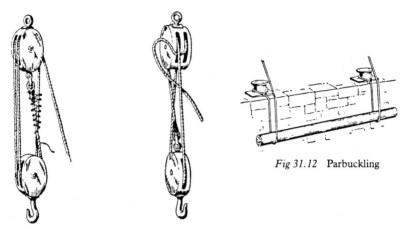

Fig 31.12 Parbuckling

Fig 31.10 Racking a tackle   *Fig 31.11* Choking a luff

Parbuckling

Parbuckling is a method by which a pole, cask or other suitably rounded object may be raised or lowered down a ramp, quayside or bulkhead. No blocks are used but a similar principle applies and as illustrated in *Fig 31.12* the weight can be lowered under control.

The mechanical advantage is 2 but there is a greater frictional loss when lifting because the ropes move around the weight itself and not through a sheave.

32   Slings and spans

Frequent causes of serious accidents are the misuse and overloading of slings and spans. The following diagrams show that the wider the angle of the slings at the lifting hook the greater the stress. Slings at a small angle share the load being lifted while slings at a wide angle share a load four times the weight being lifted.

The same applies to strops and spans where the angles between the legs should be kept as small as possible. Slings and spans should be examined before use with these facts in mind (see the following figures illustrating this principle).

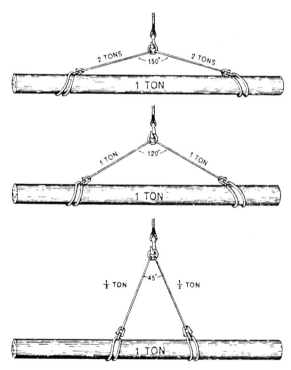

Fig 32.1 Variation of stresses in the legs of a span

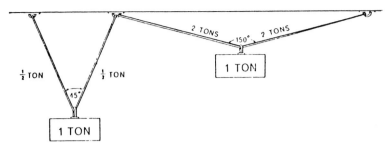

Fig 32.2 Variations of the tensions in a span.

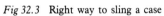

Fig 32.3 Right way to sling a case   Fig 32.4 Wrong way to sling a case

33 Bends and hitches

The rope in the diagram (*Fig 33.1*) has a whipping at one end to prevent it unlaying. One bight of the rope is 'lying Judas' (idle) in a bight on the deck and is stopped by twine to the rail. The rope is hanging in bights, while the standing part is 'hanging Judas' and ends in an eye (or bight) with the bare end seized back on the standing part.

It is dangerous to stand in the bight or to put a hand or any part of the body in a bight without first being certain that the rope – whether it be wire or fibre, large or small – cannot be worked.

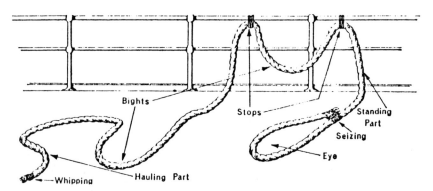

Fig 33.1 Terms used in ropework

Half-hitch, round turn, overhand knot

These two hitches and knot (*Fig 33.2*) enter into the make-up of a large number of bends and hitches.

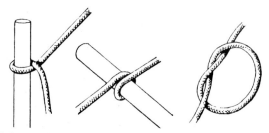

Fig 33.2 Half-hitch, round turn, overhand knot

Reef knot

The reef knot (*Fig 33.3*) consists of two overhand knots with the ends passed opposite to each other. It will not easily come undone by itself but can be untied. If it is used on ropes of unequal size or on slippery ropes the ends must be stopped to their own standing parts. If the reef knot is not formed correctly a 'granny' will result which will slip and jam and consequently be very difficult to undo when required. A seaman should not use a 'granny' knot.

Figure-of-eight knot

This knot (*Fig 33.4*) is used to prevent a rope unreeving through a block and for use with a log rotator.

Fig 33.3   Reef knot

Fig 33.4   Figure-of-eight knot

Round turn and two half-hitches, fisherman's bend

Both of these knots (*Figs 33.5* and *33.6*) are used for securing weights to a standing object, but the former is preferred as not being liable to jam. More than one round turn can be taken if desired.

Fig 33.5   Round turn and two half-hitches

Fig 33.6   Fisherman's bend

Timber hitch and half-hitch

The timber hitch (*Fig 33.7*) by itself is merely a quick means of making a running eye and is of some use on soft objects but not as useful on hard ones. With the addition of a half-hitch (*Fig 33.8*) the hold is improved. Generally speaking, the hitch and half-hitch would be used only on tapering objects with the half-hitch at the larger end. When lowering something like a spar, a length of piping or a plank into the hold, over the side or down a cliff, see that the timber hitch does not catch anywhere. It could slide towards the half-hitch which could then come off the end of the load and allow it to slip clear of the running eye.

Fig 33.7 Timber hitch   Fig 33.8 Timber hitch and a half-hitch

Clove hitch on the end, clove hitch on the bight

A most useful hitch for general purposes but it will not resist a sideways pull (*Figs 33.9* and *33.10*).

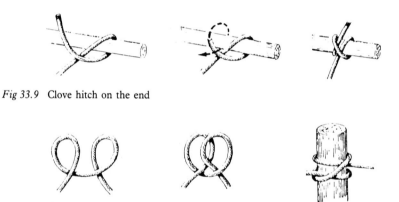

Fig 33.9 Clove hitch on the end

Fig 33.10 Clove hitch on the bight

Rolling hitch

The rolling hitch (*Fig 33.11*) is also useful for general purposes but especially where it must resist a sideways pull. Note that in the diagram, the hitch is made to resist a pull to the right.

Fig 33.11 Rolling hitch

Sheet bend or swab hitch, double sheet bend

The knot (*Fig 33.12*) is used to secure a rope's end to an eye or a small rope to a larger one.

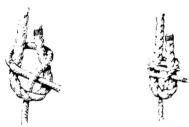

Fig 33.12 Sheet bend, double sheet bend

Fisherman's knot

The fishermen's knot (*Fig 33.13*) is used to join two smaller ropes together. When strain comes on, the two overhand knots slide together and if made as illustrated will fit snugly. The ends can then be stopped to the standing part.

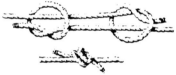

Fig 33.13 Fisherman's knot

Bowline, running bowline, bowline on the bight, French bowline

A bowline makes a reliable temporary eye, as shown in the diagram (*Fig 33.14*). A running bowline (*Fig 33.15*) makes a sliding eye. The bowline on the bight (*Fig 33.16*) can be used for lowering a man from aloft or over the side with the short bight under his arms and the long one under his seat.

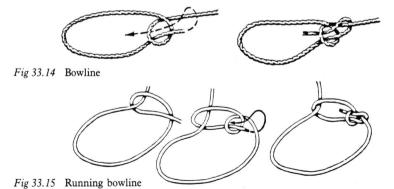

Fig 33.14 Bowline

Fig 33.15 Running bowline

276

The French bowline (*Fig 33.17*) is better for the above purpose than the bowline on the bight as the weight in the main bight keeps the arm bight taut. Form the small bight and pass the end up through it as if starting a bowline, then bring the end round (to form the arm bight) and pass it up through the small bight again and continue the bowline.

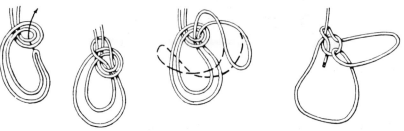

Fig 33.16   Bowline on the bight   Fig 33.17   French bowline

Butterfly knot

A party having to be roped together to climb a cliff can secure the first and last man with a bowline and each remaining man with a butterfly knot (*Fig 33.18*).

Fig 33.18   Butterfly knot

Midshipman's hitch, blackwall hitch, double blackwall hitch

All three hitches (*Figs 33.19, 33.20,* and *33.21*) are used to secure a rope to a hook. The double blackwall hitch is preferred.

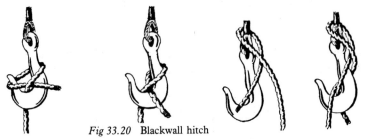

Fig 33.20   Blackwall hitch

Fig 33.19   Midshipman's hitch   Fig 33.21   Double blackwall hitch

Mousing

Mousing (*Fig 33.22*) is used to prevent a hook from unhooking.

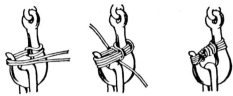

Fig 33.22 Mousing a hook

Catspaw

The catspaw (*Fig 33.23*) is used to shorten a sling.

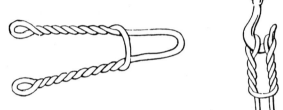

Fig 33.23 Catspaw

Marline spike hitch

This hitch (*Fig 33.24*) is used to assist in hauling taut.

Fig 33.24 Marline spike hitch, on a marline spike and on a hook

Marline hitch

The marline hitch (*Fig 33.25*) is used when lashing hammocks and similar long bundles, and as a temporary repair for chafed rope.

Fig 33.25 Marline hitch

Hawser bend

The hawser bend (*Fig 33.26*) is used to join two large ropes together. Note the seizings.

Fig 33.26 Hawser bend

Wire rope join

This diagram (*Fig 33.27*) shows the use of a wire strop in joining wire ropes.

Fig 33.27 Joining two wire hawsers with a grommet strop

Fibre rope stopper

The fibre rope stopper (*Fig 33.28*) is used to hold a fibre rope or wire rope temporarily when it is under stress.

Fig 33.28 Fibre rope stopper

Chain stopper

To make a chain stopper (*Fig 33.29*), one end of the chain is secured on to the rope using either a half-hitch or the first two parts of a rolling hitch. The bight of the hitch should be against the lay of the rope when a fibre rope is used and with the lay when a wire rope is used. The end should then be extended with the lay on a fibre rope or against the lay on a wire rope and finally stopped to the rope.

Fig 33.29 Chain stopper

Chain check stopper

The chain check stopper (*Fig 33.30*) is used to control the speed of paying out a wire.

Fig 33.30   Chain check stopper

Cod end knot

The cod end knot (*Fig 33.31*) is used to close the cod end before shooting the trawl.

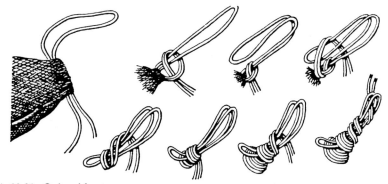

Fig 33.31   Cod end knot

34 Knots and splices

Heaving line knot

This knot (*Fig 34.1*) is for use at the end of heaving lines because it is heavy enough to carry the line the maximum distance (about 60 ft) that a man can throw. The use of weights such as steel nuts or bolts is strictly forbidden because of possible injury to persons at the other end of the throw. After the heaving line is passed, the inboard end should be bent to the hawser, which may be a mooring rope, by means of a bowline with a long eye. This will enable the eye on the hawser to be handled by the men on shore while the bowline is pulled clear of any bollard or other mooring.

Fig 34.1 Heaving line knot

Wall knot, crown knot

These knots (*Figs 34.2, 34.3* and *34.4*) form the basis of many of the knots commonly used. The crown knot by itself is used to commence a back splice.

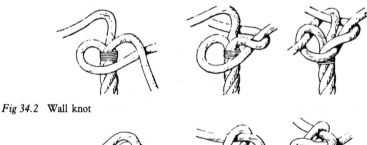

Fig 34.2 Wall knot

Fig 34.3 Crown knot

Fig 34.4 Crown and wall knot

Back splice

To make a back splice (*Fig 34.5*), first form a crown knot and tuck each strand along the rope over one laid-up strand and under the next. Two tucks of whole and one tuck of halved strands is sufficient.

Fig 34.5 Crown knot and back splice

Long splice

To make a long splice (*Fig 34.6*), unlay each rope approximately one foot for every inch in size, *ie* 2ft for a 2in rope. Marry the ropes and further unlay a strand of one and lay up a strand of the other in its place. Separate a third of the yarns of each of the two strands and tie the two-thirds portion in an overhand knot left over right. Tuck these two portions over one and under one. Carry out the same for the other end. Separate, knot and tuck the two remaining strands. Stretch the splice and cut off the yarns remaining. This splice will allow the rope to pass through a block.

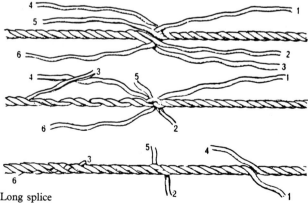

Fig 34.6 Long splice

282

Selvagee strop

The selvagee strop (*Fig 34.7*) is made of spun yarn and will grip a larger rope or wire much better than will a strop made from laid-up rope. Fix two nails at the required distance apart and pass turns with a ball of spun yarn till the strop is of the required thickness and then marl it down with marline hitches.

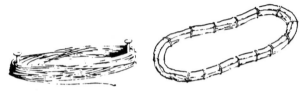

Fig 34.7 Selvagee strop

Eye splice

To make an eye splice (*Fig 34.8*), unlay the end of the rope and form a bight. Tuck the strands over one and under one. Do this three times, then stop the ends.

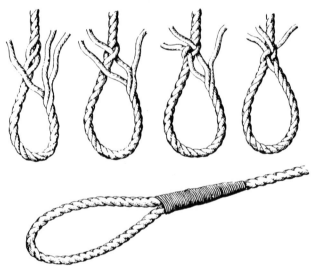

Fig 34.8 Making an eye splice

Short splice

To make a short splice (*Fig 34.9*), unlay the end of each rope and marry them together. Tuck each strand over one and under one on either end. Repeat three times.

Fig 34.9 Short splice finished by dogging the strands

Common strop, bale sling strop

A strop (*Fig 34.10*) is a length of rope with its ends spliced together. A short strop is called a common strop and is used to pass round a rope or spar *etc* to provide an eye to take a hook or shackle. A longer strop is called a bale sling strop and is used for general purposes such as slinging, hoisting, *etc*.

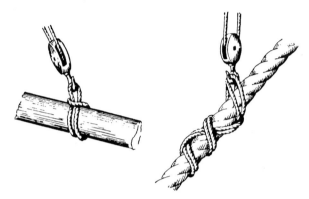

Fig 34.10 Use of a strop

Whipping

A whipping (*Figs 34.11* and *34.12*) is used on any bare end of rope in order to prevent the unlaying of the strands.

Fig 34.11 Common whipping

Fig 34.12 Bend whipping

Wire splicing

Flexible steel wire rope is six-stranded, right-handed, and has a hemp heart. Each strand also has a hemp heart. Before working wire rope good stops should be placed where the rope is to be cut and at the point to which it will be unlaid.

There are several methods for tucking the strands. The following is one example: (See *Fig 34.13*)

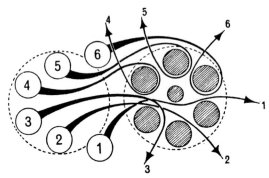

Fig 34.13 Tucking order for a wire splice

— Allow approximately 1ft of wire per inch circumference for splicing wire rope.
— Bend wire to shape the required eye and clamp in vice with the eye to the left hand side of vice if left-handed and to the right if right-handed.
— Put on a tight whipping to hold eye in position and prevent wire from unlaying.
— Take out of vice and hang up with loose wires to the left and reclamp in vice.
— Bend over wire from each strand.
— Cut off whipping at the end of wires to be tucked.
— Take out the small hearts for about 2in.
— Whip each strand tightly.
— Take the main heart between the centre of the six loose strands away from the standing wire and to the left.
— The wire nearest to you on the right is No 1 and counting clockwise from this wire they are numbered 2, 3, then main heart, 4, 5 and 6, and they are tucked in this order.
— Pass spike through the centre of the standing wire away from you, leaving the main heart and three strands on the left hand side.
— Pass No 1 wire through in the same direction as the spike, remove spike and pull wire down tight.
— Pass spike in through the same place as for No 1 but only through two wires, pass No 2 after spike and pull tight.

- Pass spike in through the same place as for No 1 but only through one wire, pass No 3 after spike and pull tight.
- Pass spike through the next wire (still working in a clockwise direction), pass wire No 4 after it, take out spike and pull tight.
- Pass spike through the next wire (still working clockwise), pass wire No 5 through and pull tight.
- Pass spike through the next wire (still working clockwise), pass wire after it, take out spike and pull tight.
- All wires have now been tucked once. Take out small hearts and cut off main heart.
- The loose wires are now run up the standing wires 'Liverpool fashion' in the following manner, working from 6 to 1.

(a) No 6 loose strand is held in left hand. Spike is then passed back through the standing wire through which the loose wire has already been passed.

(b) Loose wire No 6 is then passed through one wire under the spike and against the spike. Tighten by turning spike clockwise, then without withdrawing spike twist anti-clockwise one turn. Pass loose wire through against spike. Tighten as before. This is done until four tucks have been taken with each wire.

(c) No 5 as No 6
No 4 as No 6
No 3 as No 6
No 2 as No 6
No 1 as No 6

(d) Take off loose ends and serve.

Wire splices should be parcelled with oily canvas and served. Broken wires should not be allowed to protrude; break them off by bending backwards and forwards but do not cut.

Appendix 1

Finding the distance of objects at sea

Owing to the Earth's curvature, the distance to the sea horizon is governed by the height of the observer's eye. The figure below illustrates this and also that, in conditions of clear visibility, the distance at which the top of an object first shows itself on the horizon depends upon its height as well as the height of eye. In this case the distance of the ship from the taller of the two lighthouses is $(6 + 14\frac{1}{2}) = 20\frac{1}{2}$ miles.

HEIGHT OF EYE 8.8m
HORIZON VISIBLE 6 MILES

HEIGHT OF LIGHT 49m
HORIZON VISIBLE 14¼ MILES

HEIGHT OF LIGHT 21.3m
HORIZON VISIBLE 9¼ MILES

Range of lights

Luminous range is the maximum distance at which a light can be seen at a given time, as determined by the intensity of the light and the meteorological visibility prevailing at the time; it takes no account of elevation, observer's height of eye or the curvature of the earth.

Nominal range is the luminous range when the meteorological visibility is 10 sea miles.

Geographical range is the maximum distance at which light from a light can theoretically reach an observer, as limited only by the curvature of the earth and the refraction of the atmosphere, and by the elevation of the light and the height of the observer. Geographical ranges on charts are based on a height of eye of 15ft (5 metres).

These ranges are set out in tables at the front of the Admiralty List of Lights.

The range shown on charts published after 1972 for northern European waters is the nominal range. The range on charts published for a few years prior to 1972 was the lesser of these ranges, which, for most powerful lights in European waters would be the Geographical range. On very old charts only the Geographical range will be given.

Caution: If using old charts, geographical ranges will be given for lights not sufficiently powerful to be seen from the horizon.

All heights of lights are given above MHWS. Allowance for the state of the tides should be made with lights of small elevation.

Tables 1 and 2 allow for the effect of normal atmospheric refraction but would be of no use when conditions are abnormal.

Glare from background lighting will reduce considerably the range at which lights are sighted. A light of 100,000 candelas has a nominal range of about 20 miles; with minor background lighting as from a populated coastline this range will be reduced to about 14 miles, and with major background lighting as from a city or from harbour installations to about 9 miles.

Fishermen with a near horizon and utilising the powerful lights around the European Coastline will probably be more concerned with the Geographical Range of lights.

To find the Geographical range of a light: The height of a light is 44m and its charted visibility is 18 miles. At what distance should it be sighted if the observer's eye is 18·3m above the sea?

Distance to horizon from light 44 m (143ft) . . . . 13·7 miles

Distance to horizon from a height of 18·3m (60ft) . . . . 8·9 miles

Distance at which light should be sighted . . . . 22·6 miles

The example given refers to Tables 1 and 2.

Table 1   Distance of sea horizon in nautical miles

Height in Metres	Height in Feet	Distance in Miles	Height in Metres	Height in Feet	Distance in Miles	Height in Metres	Height in Feet	Distance in Miles	Height in Metres	Height in Feet	Distance in Miles
0·3	1	1·15	4·3	14	4·30	12·2	40	7·27	55	180	15·4
0·6	2	1·62	4·9	16	4·60	12·8	42	7·44	61	200	16·2
0·9	3	1·99	5·5	18	4·87	13·4	44	7·62	73	240	17·8
1·2	4	2·30	6·1	20	5·14	14·0	46	7·79	85	280	19·2
1·5	5	2·57	6·7	22	5·39	14·6	48	7·96	98	320	20·5
1·8	6	2·81	7·3	24	5·62	15·2	50	8·1	110	360	21·8
2·1	7	3·04	7·9	26	5·86	18	60	8·9	122	400	23·0
2·4	8	3·25	8·5	28	6·08	20	70	9·6	137	450	24·3
2·7	9	3·45	9·1	30	6·30	24	80	10·3	152	500	25·7
3·0	10	3·63	9·8	32	6·50	27	90	10·9	183	600	28·1
3·4	11	3·81	10·4	34	6·70	30	100	11·5	213	700	30·4
3·7	12	3·98	11·0	36	6·90	40	130	13·1	244	800	32·5
4·0	13	4·14	11·6	38	7·09	46	150	14·1			

Table 2   To find distance of lights rising or dipping

Height of eye													
	Metres												
	1·5	3	4·6	6·1	7·6	9·1	10·7	12·2	13·7	15·2	16·8	18·3	19·8

Height of light							Feet							
		5	10	15	20	25	30	35	40	45	50	55	60	65
m	**ft**													
12	40	9¾	11	11¾	12½	13	13½	14	14½	15	15½	15¾	16¼	16½
15	50	10¾	11¾	12½	13¼	14	14½	15	15½	15¾	16¼	16¾	17	17½
18	60	11½	12½	13½	14	14¾	15¼	15¾	16¼	16½	17	17½	17¾	18¼
21	70	12¼	13¼	14	14¾	15½	16	16½	17	17¼	17¾	18	18½	19
24	80	13	14	14¾	15½	16	16½	17	17½	18	18½	18¾	19¼	19½
27	90	13½	14½	15½	16	16¾	17¼	17¾	18¼	18½	19	19½	19¾	20¼
30	100	14	15	16	16½	17¼	17¾	18¼	18¾	19¼	19½	20	20½	20¾
34	110	14½	15¾	16½	17¼	17¾	18¼	19	19¼	19¾	20¼	20½	21	21¼
37	120	15¼	16¼	17	17¾	18¼	19	19½	20	20¼	20¾	21	21½	22
40	130	15¾	16¾	17½	18¼	19	19½	20	20½	20¾	21¼	21½	22	22½
43	140	16¼	17¼	18	18¾	19½	20	20½	21	21¼	21¾	22	22½	23
46	150	16¾	17¾	18½	19¼	19¾	20½	21	21¼	21¾	22¼	22½	23	23¼
49	160	17	18¼	19	19¾	20¼	20¾	21½	21¾	22¼	22¾	23	23½	23¾
52	170	17½	18½	19½	20	20¾	21¼	21¾	22¼	22¾	23	23½	24	24¼
55	180	18	19	20	20½	21¼	21¾	22¼	22¾	23	23½	24	24½	24¾
58	190	18½	19½	20¼	21	21½	22	22¾	23	23½	24	24½	24¾	25
61	200	18¾	20	20¾	21½	22	22½	23	23½	24	24½	24¾	25¼	25½
64	210	19¼	20¼	21	21¾	22½	23	23½	24	24½	24¾	25¼	25½	26
67	220	19½	20¾	21½	22¼	22¾	23¼	24	24¼	24¾	25¼	25½	26	26¼
70	230	20	21	22	22½	23¼	23¾	24¼	24¾	25	25½	26	26¼	26¾
73	240	20½	21½	22¼	23	23½	24	24½	25	25½	26	26¼	26¾	27
76	250	20¾	21¾	22½	23¼	24	24½	25	25½	26	26¼	26¾	27	27½
79	260	21	22¼	23	23¾	24¼	24¾	25¼	25¾	26¼	26¾	27	27½	27¾
82	270	21½	22½	23¼	24	24½	25¼	25¾	26¼	26½	27	27½	27¾	28¼
85	280	21¾	23	23¾	24½	25	25½	26	26½	27	27½	27¾	28	28½
88	290	22	23¼	24	24¾	25¼	26	26½	26¾	27¼	27¾	28	28½	28¾
91	300	22½	23½	24½	25	25¾	26¼	26¾	27¼	27½	28	28½	28¾	29¼
95	310	22¾	24	24¾	25½	26	26½	27	27½	28	28½	28¾	29	29½
98	320	23	24¼	25	25¾	26¼	27	27½	27¾	28¼	28¾	29	29½	29¾
100	330	23½	24½	25¼	26	26½	27¼	27¾	28	28½	29	29½	29¾	30
104	340	23¾	24¾	25¾	26¼	27	27½	28	28½	29	29¼	29¾	30	30½
107	350	24	25	26	26¾	27¼	27¾	28¼	28¾	29¼	29½	30	30½	30¾
122	400	25½	26½	27½	28	28¾	29¼	29¾	30¼	30¾	31	31½	32	32¼
137	450	27	28	28¾	29½	30	30¾	31¼	31¾	32	32½	33	33¼	33¾

Appendix 2

Height of tide between high and low water

A convenient table for determining the height of the tide at times between high and low water

Rise and fall of tide

Approximate rise and fall of tide at any time from high or low water

Range of Tide in Metres	0 Hour min. 20	0 Hour min. 40	1 Hour min. 0	1 Hour min. 20	1 Hour min. 40	2 Hours min. 0	2 Hours min. 20	2 Hours min. 40	3 Hours min. 0	3 Hours min. 20	3 Hours min. 40	4 Hours min. 0	4 Hours min. 20	4 Hours min. 40	5 Hours min. 0	5 Hours min. 20	5 Hours min. 40	6 Hr. min. 0
0·5	0·0	0·0	0·0	0·1	0·1	0·1	0·2	0·2	0·2	0·3	0·3	0·4	0·4	0·4	0·5	0·5	0·5	0·5
1·0	0·0	0·0	0·1	0·1	0·2	0·3	0·3	0·4	0·5	0·6	0·7	0·8	0·9	0·9	0·9	1·0	1·0	1·0
1·5	0·0	0·0	0·1	0·2	0·3	0·4	0·5	0·6	0·8	0·9	1·0	1·1	1·2	1·3	1·4	1·5	1·5	1·5
2·0	0·0	0·0	0·1	0·1	0·2	0·4	0·5	0·7	0·8	1·0	1·2	1·3	1·5	1·6	1·8	1·9	2·0	2·0
2·5	0·0	0·1	0·2	0·3	0·4	0·6	0·8	1·0	1·3	1·5	1·7	1·9	2·1	2·2	2·3	2·4	2·5	2·5
3·0	0·0	0·1	0·2	0·4	0·5	0·8	1·0	1·2	1·5	1·8	2·0	2·3	2·5	2·6	2·8	2·9	3·0	3·0
3·5	0·0	0·1	0·2	0·4	0·6	0·9	1·1	1·4	1·8	2·1	2·4	2·6	2·9	3·1	3·3	3·4	3·5	3·5
4·0	0·0	0·1	0·3	0·5	0·7	1·0	1·3	1·7	2·0	2·3	2·7	3·0	3·3	3·5	3·7	3·9	4·0	4·0
4·5	0·0	0·1	0·3	0·5	0·8	1·1	1·5	1·9	2·3	2·6	3·0	3·4	3·7	4·0	4·2	4·4	4·5	4·5
5·0	0·0	0·2	0·3	0·6	0·9	1·3	1·6	2·1	2·5	2·9	3·4	3·8	4·1	4·4	4·7	4·9	5·0	5·0
5·5	0·1	0·2	0·4	0·6	1·0	1·4	1·8	2·3	2·8	3·2	3·7	4·1	4·5	4·9	5·1	5·3	5·5	5·5
6·0	0·1	0·2	0·4	0·7	1·1	1·5	2·0	2·5	3·0	3·5	4·0	4·5	4·9	5·3	5·6	5·8	6·0	6·0
6·5	0·1	0·2	0·4	0·8	1·2	1·6	2·1	2·7	3·3	3·8	4·4	4·9	5·3	5·7	6·1	6·3	6·4	6·5
7·0	0·1	0·2	0·5	0·8	1·2	1·8	2·3	2·9	3·5	4·1	4·7	5·3	5·8	6·2	6·5	6·8	6·9	7·0
7·5	0·1	0·2	0·5	0·9	1·3	1·9	2·5	3·1	3·8	4·4	5·0	5·6	6·2	6·6	7·0	7·3	7·4	7·5
8·0	0·1	0·2	0·5	0·9	1·4	2·0	2·6	3·3	4·0	4·7	5·4	6·0	6·6	7·1	7·5	7·8	7·9	8·0
8·5	0·1	0·3	0·6	1·0	1·5	2·1	2·8	3·5	4·3	5·0	5·7	6·4	7·0	7·5	7·8	8·2	8·4	8·5
9·0	0·1	0·3	0·6	1·1	1·6	2·3	3·0	3·7	4·5	5·3	6·0	6·8	7·4	7·9	8·4	8·7	8·9	9·0
9·5	0·1	0·3	0·6	1·1	1·7	2·4	3·1	3·9	4·8	5·6	6·4	7·1	7·8	8·4	8·9	9·2	9·4	9·5
10·0	0·1	0·3	0·7	1·2	1·8	2·5	3·3	4·1	5·0	5·9	6·7	7·5	8·2	8·9	9·3	9·7	9·9	10·0
10·5	0·1	0·3	0·7	1·3	1·9	2·6	3·4	4·3	5·3	6·2	7·1	7·9	8·6	9·3	9·8	10·2	10·4	10·5
11·0	0·1	0·3	0·7	1·4	2·0	2·8	3·6	4·6	5·5	6·4	7·4	8·3	9·0	9·7	10·3	10·7	10·9	11·0
11·5	0·1	0·3	0·8	1·4	2·0	2·9	3·8	4·8	5·8	6·7	7·7	8·6	9·5	10·1	10·7	11·2	11·4	11·5
12·0	0·1	0·4	0·8	1·5	2·1	3·0	3·9	5·0	6·0	7·0	8·1	9·0	9·9	10·6	11·2	11·6	11·9	12·0
12·5	0·1	0·4	0·8	1·5	2·2	3·1	4·1	5·2	6·3	7·3	8·4	9·4	10·3	11·0	11·7	12·1	12·4	12·5
13·0	0·1	0·4	0·8	1·5	2·3	3·3	4·3	5·4	6·5	7·6	8·7	9·8	10·7	11·5	12·1	12·6	12·5	13·0
13·5	0·1	0·4	0·9	1·6	2·4	3·4	4·4	5·6	6·8	7·9	9·1	10·1	11·1	11·9	12·6	13·1	13·4	13·5
14·0	0·1	0·4	0·9	1·6	2·5	3·5	4·6	5·8	7·0	8·2	9·4	10·5	11·5	12·4	13·1	13·6	13·9	14·0
14·5	0·1	0·4	1·0	1·7	2·6	3·6	4·8	6·0	7·3	8·5	9·7	10·9	11·9	12·8	13·5	14·1	14·4	14·5
15·0	0·1	0·5	1·0	1·8	2·7	3·8	4·9	6·2	7·5	8·8	10·0	11·3	12·3	13·2	14·0	14·6	14·9	15·0

To find the *actual depth* of water at any port or place, at any given time, add the depth at Low Water to the figures given above, *opposite* the known Range of Tide for the Day, and *under* the Time required.
Reproduced from Brown's Nautical Almanac. Courtesy of Brown, Son and Ferguson Ltd.

Appendix 3

Finding distance off with the sextant

Compiled by Captain L.-G. OHMAN

				Height in metres					
Vert. Angle	1	2	3	4	5	6	7	8	9
1'	1·8562	3·7124	5·5686	7·4250	9·2812	11·137	12·994	14·850	16·706
2'	0·9281	1·8562	2·7843	3·7124	4·6405	5·5686	6·4969	7·4248	8·3530
3'	0·6187	1·2375	1·8562	2·4750	3·0937	3·7124	4·3312	4·9499	5·5686
4'	0·4640	0·9281	1·3922	1·8562	2·3203	2·7843	3·2484	3·7124	4·1765
5'	0·3712	0·7425	1·1138	1·4850	1·8562	2·2274	2·5987	2·9699	3·3412
6'	0·3094	0·6187	0·9281	1·2375	1·5469	1·8562	2·1656	2·4750	2·7843
7'	0·2652	0·5304	0·7955	1·0607	1·3259	1·5911	1·8562	2·1214	2·3866
8'	0·2320	0·4640	0·6961	0·9281	1·1601	1·3922	1·6242	1·8562	2·0883
9'	0·2062	0·4125	0·6187	0·8250	1·0313	1·2375	1·4437	1·6500	1·8562
10'	0·1856	0·3712	0·5569	0·7425	0·9281	1·1138	1·2994	1·4850	1·6706
11'	0·1687	0·3375	0·5063	0·6750	0·8437	1·0125	1·1812	1·3500	1·5187
12'	0·1547	0·3094	0·4641	0·6187	0·7734	0·9281	1·0828	1·2375	1·3922
13'	0·1428	0·2856	0·4284	0·5712	0·7139	0·8567	0·9995	1·1423	1·2851
14'	0·1326	0·2652	0·3978	0·5304	0·6629	0·7955	0·9281	1·0607	1·1933
15'	0·1238	0·2475	0·3712	0·4950	0·6187	0·7425	0·8662	0·9900	1·1137
16'	0·1160	0·2320	0·3480	0·4641	0·5801	0·6961	0·8121	0·9281	1·0441
17'	0·1092	0·2184	0·3276	0·4368	0·5460	0·6551	0·7643	0·8735	0·9603
18'	0·1031	0·2063	0·3094	0·4125	0·5156	0·6187	0·7219	0·8250	0·9281
19'	0·0977	0·1954	0·2931	0·3908	0·4885	0·5862	0·6839	0·7816	0·8793
20'	0·0928	0·1856	0·2784	0·3712	0·4641	0·5569	0·6497	0·7425	0·8353
21'	0·0884	0·1768	0·2652	0·3536	0·4420	0·5304	0·6187	0·7071	0·7955
22'	0·0844	0·1687	0·2531	0·3375	0·4219	0·5063	0·5906	0·6750	0·7594
23'	0·0807	0·1614	0·2421	0·3228	0·4035	0·4842	0·5649	0·6457	0·7263
24'	0·0773	0·1547	0·2320	0·3094	0·3867	0·4641	0·5414	0·6187	0·6961
25'	0·0742	0·1485	0·2228	0·2970	0·3712	0·4455	0·5198	0·5940	0·6682
26'	0·0714	0·1428	0·2142	0·2856	0·3570	0·4284	0·4998	0·5711	0·6425
27'	0·0687	0·1375	0·2062	0·2750	0·3437	0·4125	0·4812	0·5500	0·6187
28'	0·0663	0·1326	0·1989	0·2652	0·3315	0·3978	0·4641	0·5304	0·5966
29'	0·0640	0·1280	0·1920	0·2560	0·3200	0·3840	0·4481	0·5121	0·5761
30'	0·0619	0·1238	0·1856	0·2475	0·3094	0·3712	0·4331	0·4950	0·5569

				Height in metres				
Vert. Angle	1	2	3	4	5	6	7	8	9
31'	0·0599	0·1198	0·1796	0·2395	0·2994	0·3593	0·4191	0·4790	0·5389
32'	0·0580	0·1160	0·1740	0·2320	0·2900	0·3480	0·4060	0·4640	0·5221
33'	0·0562	0·1125	0·1687	0·2250	0·2812	0·3375	0·3937	0·4500	0·5062
34'	0·0546	0·1092	0·1638	0·2184	0·2730	0·3276	0·3822	0·4368	0·4913
35'	0·0530	0·1061	0·1591	0·2121	0·2652	0·3182	0·3712	0·4243	0·4773
36'	0·0516	0·1031	0·1547	0·2062	0·2578	0·3094	0·3609	0·4125	0·4640
37'	0·0502	0·1003	0·1505	0·2007	0·2508	0·3010	0·3512	0·4013	0·4515
38'	0·0488	0·0977	0·1465	0·1954	0·2442	0·2931	0·3419	0·3908	0·4396
39'	0·0476	0·0952	0·1428	0·1904	0·2380	0·2856	0·3332	0·3808	0·4283
40'	0·0464	0·0928	0·1392	0·1856	0·2320	0·2784	0·3248	0·3712	0·4176
41'	0·0453	0·0905	0·1358	0·1811	0·2264	0·2716	0·3169	0·3622	0·4075
42'	0·0442	0·0884	0·1326	0·1768	0·2210	0·2652	0·3094	0·3536	0·3977
43'	0·0432	0·0863	0·1295	0·1727	0·2158	0·2590	0·3022	0·3453	0·3885
44'	0·0422	0·0844	0·1266	0·1687	0·2109	0·2531	0·2953	0·3375	0·3797
45'	0·0412	0·0825	0·1237	0·1650	0·2062	0·2475	0·2887	0·3300	0·3712
46'	0·0404	0·0807	0·1211	0·1614	0·2018	0·2421	0·2825	0·3228	0·3632
47'	0·0395	0·0790	0·1185	0·1580	0·1975	0·2370	0·2764	0·3159	0·3554
48'	0·0387	0·0773	0·1160	0·1547	0·1933	0·2320	0·2707	0·3094	0·3480
49'	0·0379	0·0758	0·1136	0·1515	0·1894	0·2273	0·2652	0·3030	0·3409
50'	0·0371	0·0742	0·1114	0·1485	0·1856	0·2227	0·2599	0·2970	0·3341
51'	0·0364	0·0728	0·1092	0·1456	0·1820	0·2184	0·2548	0·2912	0·3275
52'	0·0357	0·0714	0·1071	0·1428	0·1785	0·2142	0·2499	0·2856	0·3212
53'	0·0350	0·0700	0·1051	0·1401	0·1751	0·2101	0·2451	0·2802	0·3152
54'	0·0344	0·0687	0·1031	0·1375	0·1719	0·2062	0·2406	0·2750	0·3093
55'	0·0337	0·0675	0·1012	0·1350	0·1687	0·2025	0·2362	0·2700	0·3037
56'	0·0331	0·0663	0·0994	0·1326	0·1657	0·1989	0·2320	0·2652	0·2983
57'	0·0326	0·0651	0·0977	0·1303	0·1628	0·1954	0·2279	0·2605	0·2931
58'	0·0320	0·0640	0·0960	0·1280	0·1600	0·1920	0·2240	0·2560	0·2880
59'	0·0315	0·0629	0·0944	0·1258	0·1573	0·1888	0·2202	0·2517	0·2831
60'	0·0309	0·0619	0·0928	0·1237	0·1547	0·1856	0·2165	0·2475	0·2784

Example—St. Georgio light 148 metres high   100   6·63
Vertical angle 28'   40   +2·652
  8   +0·5304

Distance 9·8   =9·8124

Additional table for vertical angles between 1°00' and 7°00'

Vertical Angle	Tabulated Angle	Vertical Angle	Tabulated Angle	Vertical Angle	Tabulated Angle	Vertical Angle	Tabulated Angle
1°10'	7'	2°40'	16'	4°10'	25'	5°40'	34'
1°20'	8'	2°50'	17'	4°20'	26'	5°50'	35'
1°30'	9'	3°00'	18'	4°30'	27'	6°00'	36'
1°40'	10'	3°10'	19'	4°40'	28'	6°10'	37'
1°50'	11'	3°20'	20'	4°50'	29'	6°20'	38'
2°00'	12'	3°30'	21'	5°00'	30'	6°30'	39'
2°10'	13'	3°40'	22'	5°10'	31'	6°40'	40'
2°20'	14'	3°50'	23'	5°20'	32'	6°50'	41'
2°30'	15'	4°00'	24'	5°30'	33'	7°00'	42'

Convert vertical angle to tabulated angle.
Use the tabulated angle in "finding distance off with the sextant".
DIVIDE THE DISTANCE TABULATED BY 10 to find the right distance.

Example—St. Georgio light 148 metres high	100	10·92
Vertical angle 2°50'	40	+ 4·368
Tabulated angle 17'	8	+ 0·8735
Distance 1·6		= 16·1615

From Brown's Nautical Almanac.
Courtesy of Brown, Son & Ferguson Ltd

Appendix 4

Specimen cards for distress transmission procedures

NAME OF SHIP .... CALL SIGN ....

DISTRESS TRANSMITTING PROCEDURES

(For use only when **IMMEDIATE ASSISTANCE** required)

1. Ensure transmitter is switched to 2,182 kHz.

2. If possible **transmit two-tone ALARM SIGNAL for $\frac{1}{2}$ to 1 minute.**

3. Then say: **MAYDAY, MAYDAY, MAYDAY**

   THIS IS......(Ship's name or call sign 3 times)......**MAYDAY**
   followed by ship's name or call sign

   POSITION ....

   NATURE OF DISTRESS ....

   AID REQUIRED .... **OVER**

4. Listen for a reply and if none heard **repeat** above procedure, particularly during the 3-minute silent period commencing at each hour and half-hour.

EXAMPLE—if possible ALARM SIGNAL followed by:

   "MAYDAY, MAYDAY, MAYDAY,

   This is NONSUCH, NONSUCH, NONSUCH,

   MAYDAY, NONSUCH,

   Position 54 25 North 016 33 West,

   I am on fire and require immediate assistance, OVER."

NOTE—(1) If language difficulties arise, use CARD 3.
(2) In providing details of aid required, the number of persons requiring aid should be given.

RECEPTION OF SAFETY MESSAGES

Any mesage which you hear prefixed by one of the following words concerns SAFETY—

MAYDAY   **PAN PAN**   **SECURITE**

(pronounced SAY-CURE-E-TAY)

If you hear these words, pay particular attention to the message and call the skipper.

MAYDAY
(Distress)   Indicates that a ship, aircraft or other vehicle is threatened by grave and imminent danger and requests immediate assistance.

PAN PAN
(Urgency)   Indicates that the calling station has a very urgent message to transmit concerning the safety of a ship, aircraft or other vehicle, or of a person.

SECURITE
(Safety)   Indicates that the station is about to transmit a message concerning the safety of navigation or giving important meteorological warnings.

Radiotelephone procedures

If language difficulties arise use *Tables 2* and *3* below, sending the word **INTERCO** to indicate that the message will be in the International Code of Signals. Call out letters as in *Table 1*. Call out numbers figure by figure as in *Table 1*.

Table 1 **Phonetic alphabet and figure-spelling tables**
(May be used when transmitting plain language or code)

Letter	Word	Pronounced as
A	Alfa	AL FAH
B	Bravo	BRAH VOH
C	Charlie	CHAR LEE or
		SHAR LEE
D	Delta	DELL TAH
E	Echo	ECK OH
F	Foxtrot	FOKS TROT
G	Golf	GOLF

H	Hotel	HOH TELL
I	India	IN DEE AH
J	Juliett	JEW LEE ETT
K	Kilo	KEY LOH
L	Lima	LEE MAH
M	Mike	MIKE
N	November	NO VEM BER
O	Oscar	OSS CAH
P	Papa	PAH PAH
Q	Quebec	KEH BECK
R	Romeo	ROW ME OH
S	Sierra	SEE AIR RAH
T	Tango	TANG GO
U	Uniform	YOU NEE FORM or OO NEE FORM
V	Victor	VIK TAH
W	Whiskey	WISS KEY
X	X-ray	ECKS RAY
Y	Yankee	YANG KEY
Z	Zulu	ZOO LOO

NB: The syllables to be emphasized are underlined.

Figure or mark to be transmitted	Word	Pronounced as
0	NADAZERO	NAH-DAH-ZAY-ROH
1	UNAONE	OO-NAH-WUN
2	BISSOTWO	BEE-SOH-TOO
3	TERRATHREE	TAY-RAH-TREE
4	KARTEFOUR	KAR-TAY-FOWER
5	PANTAFIVE	PAN-TAH-FIVE
6	SOXISIX	SOK-SEE-SIX
7	SETTESEVEN	SAY-TAY-SEVEN
8	OKTOEIGHT	OK-TOH-AIT
9	NOVENINE	NO-VAY-NINER
Decimal point	DECIMAL	DAY-SEE-MAL
Full stop	STOP	STOP

NB: Each syllable should be equally emphasized.

Table 2. **Position in code from the International Code of Signals**

(1) By bearing and distance from a landmark.
  Code letter A (Alfa) followed by a three-figure group for ship's TRUE
    bearing from landmark;
  Name of landmark;
  Code letter R (Romeo) followed by one or more figures for distance in
    nautical miles.

or

(2) By latitude and longitude.
  Latitude: Code letter L (Lima) followed by a four-figure group; (2
    figures for degrees, 2 figures for minutes) and either—N (November)
    for latitude north, or S (Sierra) for latitude south.
  Longitude: Code letter G (Golf) followed by a five-figure group; (3 figures
    for degrees, 2 figures for minutes) and either—E (Echo) for
    longitude east, or W (Whiskey) for longitude west.

Table 3. **Nature of distress in code from the International Code of
Signals**

Code Letters	Words to be transmitted	Text of Signal
AE	Alfa Echo	I must **abandon** my vessel.
BF	Bravo Foxtrot	**Aircraft is ditched** in position indicated and requires immediate assistance.
CB	Charlie Bravo	I require **immediate assistance**.
CB6	Charlie Bravo Soxisix	I require **immediate assistance**. I am on **fire**.
DX	Delta X-ray	I am **sinking**.
HW	Hotel Whiskey	I have **collided** with surface craft.
		Answer to ship in distress
CP	Charlie Papa	I am proceeding to your assistance
ED	Echo Delta	Your distress signals are understood
EL	Echo Lima	Repeat the distress position.

NB: **A more comprehensive list of signals may be found in the International Code of Signals.**

Examples of distress procedure
— Where possible, transmit ALARM SIGNAL followed by spoken words
  'Mayday Mayday Mayday'... (name of ship spoken three times or call

sign of ship spelt three times using *Table 1*) 'Mayday'. . . (name or call sign of ship) Interco Alfa Nadazero Unaone Pantafive Ushant Romeo Kartefour Nadazero Delta X-ray. '(Ship) in distress Position 015 Degrees Ushant 40 miles I am sinking.'

— Where possible, transmit ALARM SIGNAL followed by spoken words 'Mayday Mayday Mayday'. . . (name of ship spoken three times or call sign of ship spelt three times using *Table 1*) 'Mayday'. . . (name of call sign of ship) Interco Lima Pantafive Kartefour Bissotwo Pantafive November Golf Nadazero Unaone Soxisix Terrathree Terrathree Whiskey Charlie Bravo Soxisix. '(Ship) in Distress Position Latitude 54 25 North Longitude 016 33 West I am on fire.'

Appendix 5

Natural fibre ropes

3-STRAND & 8-STRAND MULTIPLAIT
to BSS 2052/77

3-STRAND DIAM. mm	GROSS WEIGHT PER 100m kg	MINIMUM STRENGTH	
		GRADE 1 MANILA kg	SISAL kg
7	3·5	370	330
8	5·4	545	483
10	6·8	705	635
12	10·5	1065	955
14	14·0	1450	1285
		tonnes	*tonnes*
16	19·0	2·03	1·80
18	22·0	2·44	2·14
20	27·5	3·25	2·85
22	33·0	3·86	3·40
24	40·0	4·57	4·07
28	53·2	6·10	5·33
32	70·0	7·90	6·86
36	89·0	9·65	8·70
40	110·0	11·94	10·42
44	134·0	14·23	12·70
48	158·5	16·77	14·74
52	187·0	19·56	17·28
56	215·0	22·35	19·82
60	248·0	25·40	22·60
64	288·0	29·00	25·70
72	362·0	35·8	32·7
80	440·0	43·5	38·7
88	535·0	51·5	46·8
96	640·0	60·0	53·5

Courtesy of H & T Marlow Ltd, Hailsham, East Sussex.

Synthetic fibre ropes

3-STRAND & 8-STRAND MULTIPLAIT
to BSS 4928 1973/74

3-STRAND	8-STRAND	NELSON Spunstaple STURDEE Split Film HARDY Monofilament POLYPROPYLENE		CORNWALL NYLON		†HERCULES H.T.POLYESTER		**SUPERMIX Mixed Polyester and Spunstaple Polypropylene	
dia. mm	Multiplait Size No.	Weight Kg/100m	Grd. min. Strength Kg	Weight Kg/100m	Grd. min. Strength Kg	Weight Kg/100m	Grd. min. Strength Kg	Weight Kg/100m	Grd. min. Strength Kg
6	—	1·70	550	2·37	750	3·00	565	—	—
8	—	3·00	960	4·20	1350	5·10	1020	4·10	1020
10	—	4·50	1425	6·50	2080	8·10	1590	6·50	1590
12	—	6·50	2030	9·40	3000	11·60	2270	9·10	2270
14	—	9·00	2790	12·80	4100	15·70	3180	12·50	3180
			tonnes		*tonnes*		*tonnes*		*tonnes*
16	—	11·50	3·5	16·60	5·3	20·50	4·1	15·40	4·1
18	1	14·80	4·5	21·00	6·7	26·00	5·1	18·50	5·1
20	2½	18·00	5·4	26·00	8·3	32·00	6·3	24·00	6·3
22	—	22·00	6·5	31·50	10·0	38·40	7·6	28·40	7·6
24	3	26·00	7·6	37·50	12·0	46·00	9·1	34·00	9·1
28	3½	35·50	10·1	51·00	15·8	63·00	12·2	47·00	12·2
32	4	46·00	12·8	66·50	20·0	82·00	15·7	61·00	15·7
36	4½	58·50	16·1	84·00	24·8	104·00	19·3	76·00	19·3
40	5	72·00	19·4	104·00	30·0	128·00	27·5	95·00	23·9
44	5½	88·00	23·4	126·00	35·8	155·00	32·7	115·00	28·4
48	6	104·00	27·2	150·00	42·0	185·00	38·5	136·00	33·5
52	6½	122·00	31·5	175·00	48·8	215·00	45·0	160·00	39·1
56	7	142·00	36·0	203·00	56·0	251·00	51·4	185·00	44·7

(continued)

3-STRAND dia. mm	8-STRAND Multiplait Size No.	NELSON Spunstaple STURDEE Split Film HARDY Monofilament POLYPROPYLENE		CORNWALL NYLON		†HERCULES H.T.POLYESTER		**SUPERMIX Mixed Polyester and Spunstaple Polypropylene	
		Weight Kg/100m	Gtd. min. Strength Kg	Weight Kg/100m	Gtd. min. Strength Kg	Weight Kg/100m	Gtd. min. Strength Kg	Weight Kg/100m	Gtd. min. Strength Kg
60	7½	163·00	41·2	233·00	63·9	288·00	57·3	212·00	49·8
64	8	185·00	46·6	265·00	72·0	328·00	67·0	242·00	57·9
72	9	234·00	58·5	336·00	90·0	415·00	83·0	306·00	72·1
80	10	290·00	72·0	415·00	110·0	512·00	103·0	378·00	88·4
88	11	351·00	86·4	502·00	131·0	614·00	122·0	457·00	106·0
96	12	417·00	102·0	598·00	154·0	736·00	144·0	545·00	125·0
104	13	490·00	120·0	700·00	180·0	860·00	167·0	640·00	145·0
112	14	570·00	135·0	810·00	210·0	1000·00	189·0	740·00	165·0
120	15	650·00	155·0	930·00	240·0	1150·00	219·0	850·00	190·0
128	16	740·00	175·0	1060·00	270·0	1310·00	247·0	970·00	215·0
144	18	940·00	220·0	1340·00	340·0	1660·00	311·0	1220·00	270·0
160	20	1160·00	270·0	1660·00	420·0	2040·00	390·0	1510·00	330·0
168	21	1280·00	300·0	1830·00	460·0	2250·00	430·0	1660·00	365·0
176	22	1400·00	325·0	2010·00	510·0	2470·00	470·0	1830·00	400·0
184	23	1530·00	360·0	2190·00	560·0	2700·00	500·0	2000·00	440·0
192	24	1670·00	390·0	2390·00	620·0	2930·00	550·0	2180·00	470·0

† HERCULES POLYESTER – Sizes up to 36mm dia. to BSS 4928
Sizes 40mm dia. and over to Company specification (strength exceeds British Standards).
** Manufactured to Company specification.

Courtesy of H & T Marlow Ltd, Hailsham, East Sussex

Tables of weights and measures

Nautical measures

1 ton (displacement) = 35 cu. feet of Salt Water or 36 cu. feet of Fresh Water
1 ton (register) = 100 cu. feet   **1 ton (measurement)** = 40 cu. feet

Fresh Water

1 cubic foot = $6\frac{1}{4}$ gallons and weighs 62·5 lb (1000 oz.)
36 cubic feet = 224 gallons and weights 1 ton
1 gallon = 4·536 litres and weighs 10 lb
10 British gallons = 12 American gallons (approx.)

Salt Water

1 cubic foot weighs 64 lb
35 cubic feet weigh 1 ton = Approx 218 gallons (depending on density)

Shackles of cable to metres

Shackles	Metres		Shackles	Metres
1 = 15 fathoms =	27·432		6 = 90 fathoms =	165 approx.
2 = 30 fathoms =	55 approx.		7 = 105 fathoms =	192 approx.
3 = 45 fathoms =	82 approx.		8 = 120 fathoms =	220 approx.
4 = 60 fathoms =	110 approx.		9 = 135 fathoms =	246 approx.
5 = 75 fathoms =	140 approx.		10 = 150 fathoms =	274 approx.

Lengths in British units

12 inches = 1 foot   3 feet = 1 yard   6 feet = 1 fathom
1 cable = 608 feet (which is roughly 100 fathoms or 200 yards).
10 cables = 1 nautical or 1 sea mile.
1 nautical mile = 6080 feet (roughly 2000 yards) = 1·15 statute miles.
1 statute mile (or land mile) = 5280 feet = 1760 yards = 0·87 sea miles.

Other measures of weight

1 pig of ballast = 56 lb. ∴ 2 pigs = 1 cwt. and 40 pigs = 1 ton
1 long ton (British) = 2,240 lb. = 1·12 short tons = 1·016 metric tonnes.

Feet to metres 1ft = 0·3048 m.

Feet	Metres	Feet	Metres	Feet	Metres	Feet	Metres
1	0·30	26	7·92	51	15·54	76	23·16
2	0·61	27	8·23	52	15·85	77	23·47
3	0·91	28	8·53	53	16·15	78	23·77
4	1·22	29	8·84	54	16·46	79	24·08
5	1·52	30	9·14	55	16·76	80	24·38
6	1·83	31	9·45	56	17·07	81	24·69
7	2·13	32	9·75	57	17·37	82	24·99
8	2·44	33	10·06	58	17·68	83	25·30
9	2·74	34	10·36	59	17·98	84	25·60
10	3·05	35	10·67	60	18·29	85	25·91
11	3·35	36	10·97	61	18·59	86	26·21
12	3·66	37	11·28	62	18·90	87	26·52
13	3·96	38	11·58	63	19·20	88	26·82
14	4·27	39	11·89	64	19·51	89	27·13
15	4·57	40	12·19	65	19·81	90	27·42
16	4·88	41	12·50	66	20·12	91	27·74
17	5·18	42	12·80	67	20·42	92	28·04
18	5·49	43	13·11	68	20·73	93	28·35
19	5·79	44	13·41	69	21·03	94	28·65
20	6·10	45	13·72	70	21·34	95	28·96
21	6·40	46	14·02	71	21·64	96	29·26
22	6·71	47	14·33	72	21·95	97	29·57
23	7·01	48	14·63	73	22·25	98	29·87
24	7·32	49	14·94	74	22·56	99	30·18
25	7·62	50	15·24	75	22·86	100	30·48

Metres to feet 1 metre = 3·2808 ft.

Metres	Feet	Metres	Feet	Metres	Feet	Metres	Feet
1	3·28	26	85·30	51	167·32	76	249·34
2	6·56	27	88·58	52	170·60	77	252·62
3	9·84	28	91·86	53	173·88	78	255·91
4	13·12	29	95·14	54	177·17	79	259·19
5	16·40	30	98·43	55	180·45	80	262·47
6	19·69	31	101·71	56	183·73	81	265·75
7	22·97	32	104·99	57	187·01	82	269·03
8	26·25	33	108·27	58	190·29	83	272·31
9	29·53	34	111·55	59	193·57	84	275·59
10	32·81	35	114·83	60	196·86	85	278·87
11	36·09	36	118·11	61	200·13	86	282·15
12	39·37	37	121·39	62	203·41	87	285·43
13	42·65	38	124·67	63	206·69	88	288·71
14	45·93	39	127·95	64	209·97	89	291·99
15	49·21	40	131·23	65	213·25	90	295·28
16	52·49	41	134·51	66	216·54	91	298·56
17	55·77	42	137·80	67	219·82	92	301·84
18	59·06	43	141·08	68	223·10	93	305·12
19	62·34	44	144·36	69	226·38	94	308·40
20	65·62	45	147·64	70	229·66	95	311·68
21	68·90	46	150·92	71	232·94	96	314·96
22	72·18	47	154·20	72	236·22	97	318·24
23	75·46	48	157·48	73	239·50	98	321·52
24	78·74	49	160·76	74	242·78	99	324·80
25	82·02	50	164·04	75	246·06	100	328·08

List of useful publications

Admiralty publications

The Mariner's Handbook
Symbols and Abbreviations used on Admiralty Charts
Admiralty List of Lights (Volume covering required area)
Admiralty Tide Tables (Volume covering required area)
Admiralty Pilot or Sailing Directions (Volume for area)
Admiralty Notices to Mariners and Annual Summary

International Maritime Organisation (IMO)

Torremolinos International Convention for the Safety of Fishing Vessels 1977
Basic Principles and Operational Guidance relating to Navigational Watch-keeping

UK Department of Transport

The Fishing Vessels (Safety Provisions) Rules
Merchant Shipping Notices
Recommended Code of Safety for Fishermen
Fires in Ships
Personal Survival at Sea

Tables and Almanacs

Burton's or Norie's Tables
Brown's or Reed's Nautical Almanac

See also the list of Fishing News Books publications at end of book

Brief resume of insurance cover in fishing vessel policy issued by the British Marine Mutual Insurance Association Limited

Hull and engines of insured ship

— Damage caused by 'perils of the sea'
— Damage caused by riots or labour disturbances.
— Damage caused by breakage of machinery parts (even if the original breakage was not caused by peril of the sea), but excluding parts replaced because of inherent defect or ordinary wear and tear.
— Salvage charges, sue and labour expenses and general average all incurred in avoiding or reducing claims.

Liability to others — Protection and indemnity (P&I)

— Liability to other ships, to persons aboard other ships and to cargoes for damage caused by negligence.
— Liability for damage to fixed property, including pollution.
— Liability under towage contracts.
— Liability for wreck removal.
— Employers liability to crew (but only on request) and expenses which the owner is legally liable for arising from personal death illness or injury.
— The costs of defending any of the above claims.

General exclusions and provisos to all claims

— No claims are recoverable if arising from recklessness of the assured, from unseaworthiness of the vessel, or from deliberate acts subjecting the vessel to dangers outside the scope of those hazards ordinarily and reasonably arising in the course of the owner's business, or attributable to the wilful misconduct of the owner.
— No claim shall exceed the sum insured and it is assumed that the owner will limit his liability if he is able to do so.
— No claims shall be paid for loss of market, delay, detention or demurrage.
— The Association has rights of subrogation after paying any claim.

Index

International Code of Signals

The meanings of all single letter flags, A to Z,
are given on page 175

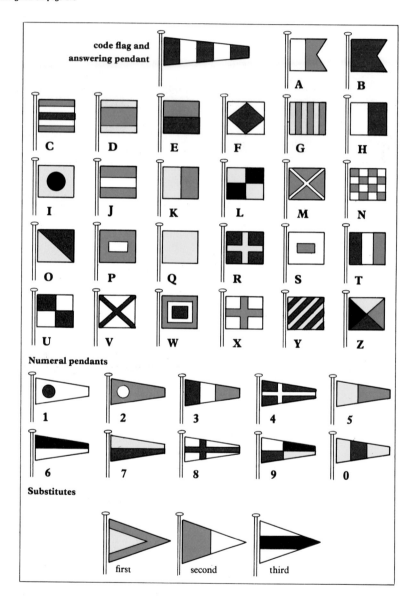

Marks in Region A (see page 36)

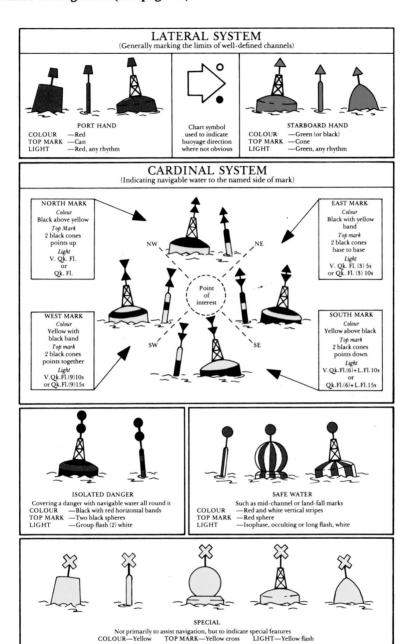

LATERAL SYSTEM
(Generally marking the limits of well-defined channels)

PORT HAND
COLOUR —Red
TOP MARK —Can
LIGHT —Red, any rhythm

Chart symbol used to indicate buoyage direction where not obvious

STARBOARD HAND
COLOUR —Green (or black)
TOP MARK —Cone
LIGHT —Green, any rhythm

CARDINAL SYSTEM
(Indicating navigable water to the named side of mark)

NORTH MARK
Colour Black above yellow
Top Mark 2 black cones points up
Light V. Qk. Fl. or Qk. Fl.

EAST MARK
Colour Black with yellow band
Top mark 2 black cones base to base
Light V. Qk. Fl. (3) 5s or Qk. Fl. (3) 10s

WEST MARK
Colour Yellow with black band
Top mark 2 black cones points together
Light V.Qk.Fl.(9)10s or Qk.Fl.(9)15s

SOUTH MARK
Colour Yellow above black
Top mark 2 black cones points down
Light V.Qk.Fl.(6)+L.Fl.10s or Qk.Fl.(6)+L.Fl.15s

Point of interest

ISOLATED DANGER
Covering a danger with navigable water all round it
COLOUR —Black with red horizontal bands
TOP MARK —Two black spheres
LIGHT —Group flash (2) white

SAFE WATER
Such as mid-channel or land-fall marks
COLOUR —Red and white vertical stripes
TOP MARK —Red sphere
LIGHT —Isophase, occulting or long flash, white

SPECIAL
Not primarily to assist navigation, but to indicate special features
COLOUR—Yellow   TOP MARK—Yellow cross   LIGHT—Yellow flash

Marks in Region B (see page 36)

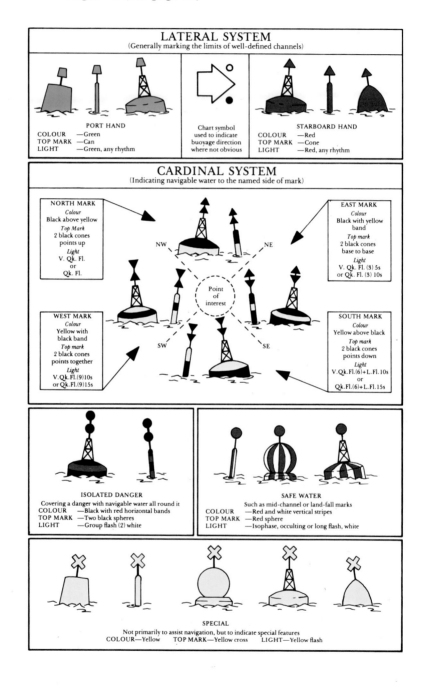

LATERAL SYSTEM
(Generally marking the limits of well-defined channels)

PORT HAND
COLOUR —Green
TOP MARK —Can
LIGHT —Green, any rhythm

Chart symbol used to indicate buoyage direction where not obvious

STARBOARD HAND
COLOUR —Red
TOP MARK —Cone
LIGHT —Red, any rhythm

CARDINAL SYSTEM
(Indicating navigable water to the named side of mark)

NORTH MARK
Colour Black above yellow
Top Mark 2 black cones points up
Light V. Qk. Fl. or Qk. Fl.

EAST MARK
Colour Black with yellow band
Top mark 2 black cones base to base
Light V. Qk. Fl. (3) 5s or Qk. Fl. (3) 10s

WEST MARK
Colour Yellow with black band
Top mark 2 black cones points together
Light V.Qk.Fl.(9)10s or Qk.Fl.(9)15s

SOUTH MARK
Colour Yellow above black
Top mark 2 black cones points down
Light V.Qk.Fl.(6)+L.Fl.10s or Qk.Fl.(6)+L.Fl.15s

Point of interest

NW NE SW SE

ISOLATED DANGER
Covering a danger with navigable water all round it
COLOUR —Black with red horizontal bands
TOP MARK —Two black spheres
LIGHT —Group flash (2) white

SAFE WATER
Such as mid-channel or land-fall marks
COLOUR —Red and white vertical stripes
TOP MARK —Red sphere
LIGHT —Isophase, occulting or long flash, white

SPECIAL
Not primarily to assist navigation, but to indicate special features
COLOUR—Yellow TOP MARK—Yellow cross LIGHT—Yellow flash

IALA maritime buoyage system, Region A

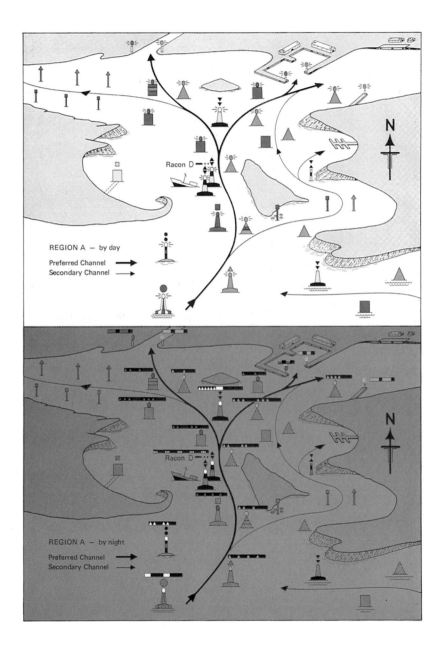

IALA maritime buoyage system, Region B

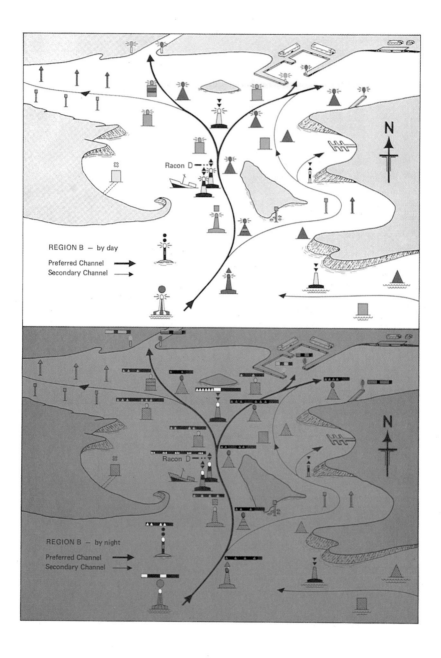

Collision Avoidance Regulations — Light and shapes to be shown

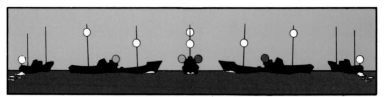

RULE 23(a). Power-driven vessel under way.

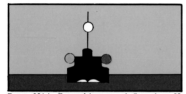

RULE 23(a). Power-driven vessel (less than 50 metres) under way.

RULE 23(a). Power driven vessel (less than 50 metres) under way.

RULE 23(b). Air-cushion vessel in non-displacement mode.

RULE 24(a) & (d). Power-driven vessel towing by day and vessel towed. Length of tow more than 200 metres.

RULE 24(a) & (e). Power-driven vessel towing. Length of tow 200 metres or less.

RULE 24(a) & (e). Power-driven vessel towing. Length of tow more than 200 metres.

RULE 24(a) & (g). Power-driven vessel towing by day, an inconspicuous object. Length of tow less than 200 metres

RULE 25(a). Sailing vessel under way.

RULE 25(c). Sailing vessel under way carrying optional lights.

RULE 25(e). Vessel under sail and power by day. (See also Rule 3(b) and (c).)

318

RULE 25(d) (i) or (ii). Small vessel not power-driven and RULE 23(c) small slow power-driven vessels.

RULE 26(b). Vessel trawling by day.

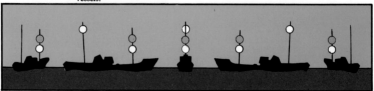

RULE 26(b). Vessel trawling but not making way.

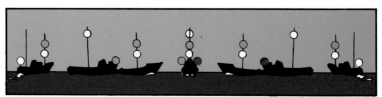

RULE 26(b). Vessel trawling and making way.

RULE 26(c). Vessel fishing but not trawling. Gear extends horizontally more than 150 metres.

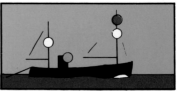

RULE 26(c). Vessel fishing, making way, but not trawling. Gear extends horizontally more than 150 metres.

RULE 26(c). Vessel fishing by day, but not trawling. Gear extends horizontally more than 150 metres.

RULE 26(c). Vessel fishing, but not trawling. Gear extends horizontally 150 metres or less.

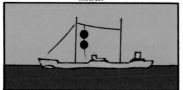

RULE 27(a). Vessel not under command by day.

RULE 27(a). Vessel not under command, not making way.